ADVANCED SIGNAL PROCESSING FOR COMMUNICATION SYSTEMS

THE KLUWER INTERNATIONAL SERIES
IN ENGINEERING AND COMPUTER SCIENCE

ADVANCED SIGNAL PROCESSING FOR COMMUNICATION SYSTEMS

edited by

Tadeusz A. Wysocki
University of Wollongong, Australia

Michael Darnell
The University of Leeds, United Kingdom

Bahram Honary
Lancaster University, United Kingdom

SPRINGER SCIENCE+BUSINESS MEDIA, LLC

Library of Congress Cataloging-in-Publication Data

International Symposium on Digital Signal Processing for Communication Systems (6th : 2002 : Sydney, Australia)
Advanced signal processing for communication systems / edited by Tadeusz A. Wysocki, Michael Darnell, Bahram Honary.
p. cm.—(The Kluwer international series in engineering and computer science ; SECS 703)
Includes bibliographical references and index.
ISBN 978-1-4757-7620-1 ISBN 978-0-306-47791-1 (eBook)
DOI 10.1007/978-0-306-47791-1
1. Signal processing--Congresses. I. Wysocki, Tadeusz. II. Darnell, Mike. III. Honary, Bahram. IV. Title. V. Series.

TK5102.9 .I548 2002
621.382'2 21

2002028516

Originally published by Kluwer Academic Publishers in 2002
Softcover reprint of the hardcover 1st edition 2002

Printed on acid-free paper.

CONTENTS

PREFACE

In the second year of the twenty first century, we are witnessing unprecedented growth in both quality and quantity of services offered by communication systems. Most of the recent advancements in communication systems performance have been only made possible by application of digital signal processing in all areas of communication systems development and implementation. Advanced digital signal processing allows for the new generation of communication systems to approach the theoretical predictions, and to practically utilize the ideas that have not been considered feasible to implement not so long ago. This book consists of 20 selected and revised papers from the 6th International Symposium on Digital Signal Processing for Communication Systems, held in January 2002, at Pacific Parkroyal Hotel in Manly, Sydney, Australia.

The first group of papers, deals with the audio and video processing for communications applications, and includes topics ranging from multimedia content delivery over the Internet, through the speech processing and recognition to recognition of non-speech sounds that can be attributed to the surrounding environment.

Another theme which receives significant attention in this book is orthogonal freqency division multiplexing (OFDM) in its various forms, eg HIPERLAN, IEEE 802, 11 a. Aspects of OFDM technology, which are covered here, include novel forms of modulation and coding, methods of reducing in-band and out-of-band spurious signal generation, and means of reducing the peak-to-average power ratio of an OFDM waveform. In these contributions, a key objective is to return the inherent implementational simplicity of the OFDM technique whilst enhancing its performance relative to single carrier systems.

Digital signal processing for second and third generation systems is represented in the book as well. The topics covered here include both theoretical issues like spreading sequence design and implementation issues of 3G user equipment modem, and MMSE receivers for CDMA systems. A useful comparison of complexity of channel estimation, equalization and decoding for GSM receivers is discussed, too.

The book also includes useful papers on applications of error control coding and information theory. These start with mathematical structure and decoding techniques and continue with channel capacity approaching codes and their applications to various communication systems.

The last group of papers included in the book consider several important issues of digital signal processing for communication systems like modulation, software defined radio, and channel estimation.

The Symposium was made possible by the generous support of the New South Wales Section of IEEE, the Smart Internet Technology Cooperative Research Center, the Telecommunications and Information Technology Research Institute, the Australian Telecommunications Cooperative Research Center, and the School of Electrical, Computer, and Telecommunications Engineering at the University of Wollongong. The Organizing Committee is most grateful for their support. The editors wish to thank the authors for their dedication and lot of efforts in preparing their contributions, revising and submitting their chapters as well as everyone else who participated in preparation of this book.

Tadeusz Wysocki
Mike Darnell
Bahram Honary

Chapter 1

APPLICATION OF STREAMING MEDIA IN EDUCATIONAL ENVIRONMENTS

Parviz Doulai
Educational Delivery Technology Laboratory (EDTLab), University of Wollongong, Wollongong NSW 2500, Australia

Abstract: This paper discusses the growing application of Web-based instruction and examines real time streaming technology in educational settings. The steps required in the process of applying streaming technology in education are outlined, and available tools and the nature of the delivery platforms are identified. The prospects and challenges in introducing virtual learning environments to tertiary institutions are illustrated using two case studies. It will not be long to overcome the challenges confronting technology-based education in traditional teaching institutions.

Key words: Educational Technology, Streaming, Multimedia, Virtual Learning Environment, Virtual Classroom

1. INTRODUCTION AND BACKGROUND

Educational institutions have long been a testing ground for the latest technological breakthroughs that change the way professional educators work and live. Examples include the growing application of information communications technologies and the use of network delivered multimedia educational modules through the application of interactive and dynamic Web environments. The growing global information technology revolution has already changed the face and culture of teaching and learning in Australia and other parts of the world, creating new opportunities and challenges for professional educators. The new and emerging educational technologies have enabled academic institutions to provide a flexible and more open learning environment for students. It is shown that in a well-designed web-based support system, students take more responsibility for their own learning, and instructors function more like coaches and mentors for a new generation of professionals [1].

The outcome of research and development work in utilizing new and emerging educational technologies in traditional educational institutions has also found its way in serving distance students. The convergence of new information technologies such as telecommunications, computers, satellites, and fiber optic technologies is making it easier for teaching institutions to implement distance education [2,3]. National and transnational virtual universities as well as traditional educational establishments are offering online degree programs, continuing education and corporate training courses. In many cases Web-based instruction and course management tools are used to deliver courseware containing interactive multimedia-based educational modules.

An integrated environment containing Web-based course delivery and management along with multimedia modules is commonly referred to as a virtual learning environment or a virtual classroom. Virtual learning environments are used to support real classroom environments in traditional academic institutions [4,5]. Virtual classrooms also were found to be very attractive in virtual campuses and virtual universities all around the globe [6]. Key technologies involved in the development of virtual learning environments include multimedia and streaming media.

Reasons for developing and utilizing virtual classrooms by teaching institutions vary, some endeavor to keep up with the ever changing frontiers of educational technologies, whilst others see it as an approach that gives students more control over their learning. The use of new and emerging educational technologies offers students a dynamic learning environment through which class communication and collaboration can be achieved with minimum time and budget requirements. In fact, the great benefit of online learning in general and virtual classrooms in particular is that it provides educators with an opportunity to get students to collaborate and to communicate very easily [1]. Two key issues in online learning are retention and the development of interactive and collaborative activities and environments. Creating a motivational and interactive virtual learning environment can enhance student retention, completion, and overall enthusiasm for this new type of learning arena [1].

In applications related to online learning, multimedia is the ability to include sound and video into Web pages. Due to the availability of many public domain and commercial computer programs it has become increasingly easy to incorporate audio and video clips into any digital document or multimedia Web publishing materials. Streaming media came about in response to the problem of bandwidth-greedy multimedia files, opening the possibilities of delivering many multimedia applications via the Internet. Streaming refers to the process of delivering audio clips, video clips, and other media in real-time to online users [7].

Streamed audio and video files can be found in a number of World Wide Web locations serving a wide-variety of purposes, such as a vocal introduction to a homepage, a movie trailer, or an interactive educational presentation. One of the major attractions to streaming media is "live" broadcasting that has less applicability to educational environment. In a simple educational setup, the streaming media is used to deliver synchronized text, images and other media files over the public TCP/IP network. In a more complex setup, streaming is used for network delivery of interactive multimedia modules [8].

This paper illustrates two case studies; a simple virtual classroom offering standard Power Point slides synchronized with streamed voice narration and a stream video presentation in which the video is indexed to the table of the content. These case studies are explained in terms of the module structure and the method of delivery. Both modules are delivered to students over a low bandwidth modem connection.

2. STREAMING: MULTIMEDIA FILES FOR NETWORK DELIVERY

It would be useful to utilize desktop videos for course material presentation and distribution. However, until recent times network delivery of multimedia clips was limited to a corporate environment or on-campus environment where students have direct access to high-speed lines. The delivery of media files over the Web has always been limited by the bandwidth of communication lines or channels. Development in this field is happening in two directions: faster connections and communication technologies [9] that are altering the capacity of the communication channels and new multimedia technologies for the Web, such as streaming audio and video, flash animation, and others that are allowing for better delivery of media on the Web [4].

When video first came to the World Wide Web, it was necessary to download the entire video file before it could be played. This was seen to be one major disadvantage of traditional multimedia clips and modules. Downloading typically megabytes of video files resulted in substantial delays before the audience could actually hear or view the clip. This was even worse when large clips were downloaded over a slow modem connection.

Streaming media is a method of providing audio, video and other media files in real-time without download waits over the Internet or corporate Intranet. Instead of downloading the file in its entirety before playing it, streaming technology takes a different approach; it downloads the beginning of the file, forms a buffer of packets, and when an appropriate buffer is reached, the client player plays back the packets in a seamless stream. While the viewer is watching, it downloads the next portion, etc., until the entire file is played. The buffer provides a way for the player to protect itself in case of network congestion, lost packets, or other interference.

2.1 History: Streaming Audio and Video

Progressive Networks [10] led the way in the development of streaming audio and video, launching "RealAudio 1.0" in 1995. "RealAudio 2.0" was then announced that upgraded sound to "FM mono" quality and made live Webcasting possible for the first time. RealAudio 2.0 introduced important features such as server bandwidth negotiation, support for firewalls and open Application Programming Interface (API) for third party developers. Compatibility of RealAudio 2.0 with the Netscape Navigator plug-in architecture made it possible to play RealAudio content available as an integrated part of a Web page. In February 1997, Progressive Networks released RealVideo 1.0 that made delivery of video over 28.8 kbps a reality. The system also offered full-motion-quality video using V.56 (56kbps) and near TV broadcast quality video at Local Area Network (LAN) rates or broadband speeds (100 kbps and above).

In October 1997, Progressive Networks officially changed its name to Real Networks prior to the release of what it called "RealSystem 5.0". The system included RealPlayer 5.0, RealEncoder 5.0, RealServer 5.0 and a software called RealPublisher. Until the release of RealSystem 6.0 in 1999, the delivery of multimedia files were conducted using Real Networks propriety PNM (Progressive Networks Metafile) format. RealSystem 6.0 used the Real Time Streaming Protocol (RTSP) that was then a new standard for improved server-client communication. RealSystem 6.0 could also stream and play not just Real Networks own format, but also standard data types such as MIDI, AVI or QuickTime. Case studies illustrated in this paper were based on RealSystem 6.0.

Real Time Streaming Protocol is designed to work with time-based media, such as streaming audio and video, as well as any application where application-controlled, time-based delivery is essential. In addition, RTSP is designed to control multicast delivery of streams, and is ideally suited to full multicast solutions [7]. Currently, RealSystem supports a variety of new data types. These include audio and video as well as text, images and animation.

In fact, streaming now is seen to be a platform for delivering information, rather than just as a system for delivering video. One can tie other kinds of Web content to the timeline of a video or an audio presentation. This allows the creation of a complex and personalized experiences for the end user. An example that contains a variety of media files with precise timing structure is available in [11].

2.2 Why Streaming?

There are several reasons why downloading of an entire media file prior to its play back is unsuitable in the delivery of information over the public TCP/IP network. For instance, if a user on a low bandwidth connection (and even high bandwidth) wants to move forward in the video they have to wait until the whole file is downloaded. Also, if a user only views a small portion of the stream and they are on a high bandwidth connection they are likely to have downloaded the whole file after only a few seconds. This will cost the user extra bandwidth because Web servers typically download as fast as they can. Moreover, Web severs do not have Intellectual Property control and so a publisher will not be able to prevent users from downloading the media file for re-using. Also Web servers are not capable of delivering presentations of unlimited or undetermined length, as well as live broadcast of media files. There are other reasons as well, which proves the superiority of dedicated streaming servers over the standard web servers in the delivery of multimedia files.

Streaming multimedia has been optimized for use on the Internet in two ways:

- Clips are highly compressed, so that download time is drastically reduced. The goal is to download the clip faster than it takes to play the clip, even when using dial up modem connections.
- The players and plug-ins can play the clip as it is being downloaded. They start playing immediately, thus reducing wait time for the user.

These optimizations allow users to do things that are impractical for traditional multimedia including broadcasting of live audio and video events and broadcasting of extremely large multimedia files, such as audio books that can take many hours to play. Often delivery of multimedia files through a dedicated stream server is combined with fast-forward and rewind capabilities.

3. STREAMING MEDIA: SERVERS, PLAYERS AND ENCODERS

RealSystem, Microsoft Windows Media Technologies [12] and Apple's QuickTime [13] offer tools for streaming multimedia content across corporate Intranets and the Internet. They allow the use of scripting languages to control the player or more importantly the integration with the browser so that one can embed the player and control it using Java script. Exposure to Java is useful as it ensures the developers can use the wealth of Java in virtual classrooms.

Producing a pre-recorded streaming multimedia requires the following steps:

1. Recording the content that requires proper recording equipment such as video cameras, microphones, etc.
2. Digitization or conversions of resulting clip into a multimedia format, such as .wav, .avi, .mov, rm, etc. It is possible to do this at the same time as step one by recording directly to the multimedia format.
3. Post-processing in the multimedia format, such as adjusting sound quality, editing the content, etc.
4. Conversions of the resulting multimedia format into a preferred format (eg. RealSystem format) using the relevant encoder (eg. RealProducer). If there are no editing enhancements, one can record direct to the preferred format.
5. Uploading the resulting file on a Web server, or a dedicated steaming server such as RealServer, so people on the Web can access it as streaming multimedia.

Examples shown in this paper use RealSystem, which is a collection of components by Real Networks for producing and distributing streaming multimedia. The three components of RealSystem include:

- Producer Module (encoder) that converts existing multimedia files into RealSystem format. The encoder program can also record to RealSystem format directly from audio and video sources.
- Player Module that plays, amongst other things, the RealSystem media file formats. The free version of RealPlayer includes both as an external version, and a Web browser plug-in version. The professional version of RealPlayer adds the ability to record broadcasts and other advanced features.
- Server Module that offers live broadcast and advanced features like automatic adjustments of transmission speeds to match user's connection, or the ability to fast forward and rewind.

4. VIRTUAL LEARNING ENVIRONMENTS

Web-based instruction can be supplemented by audio and video files to closely simulate a real classroom environment. Streaming technology is the key technology used in delivery of educational multimedia modules over the network. A virtual learning environment in its relatively complete form contains a small size video clips that shows the class activity as well as a series of text pages and images representing the content of the blackboard and the overhead projector screen.

From a developer of educational resources perspectives, the interesting idea behind streaming files is the synchronization of the playback of arbitrary files such as text, images etc. For instance, one can synchronize a flash animation file with an audio, text, image, or any other data files. In a virtual classroom environment, one can synchronize the playback of a class video with images taken from the blackboard or the overhead screen as the lecture progresses.

4.1 Case Study 1: Stream Video Integration into Virtual Classrooms

Due to the recent availability of video compressor/decompressor (codec) technologies with compressions designed for web delivery, it is now possible to use video as an effective resource in a web-based instruction environment. Different client programs are now available to make movies with different data rates, and different streaming server programs are now available to negotiate with the client machines to deliver stream video at relatively high quality even via narrow bandwidth of modem connections.

A stream video presentation was included into a combined final year and Master subject (ELEC476/912) learning environment to provide background materials for students group projects. This module was offered in two formats to meet low- and high-end Mac, PC and UNIX platforms as well as slow and moderately fast network connections. In both formats an audio and a video file synchronized with text and images were used to create a simple virtual tutorial classroom.

An interesting feature of most streaming server programs is that they allow client machines to directly negotiate with the server to access the part on the media file it wants. Normally, after a short pause the user can jump to anywhere in an audio or video clip. The video can be indexed to a table of contents and can also automatically "flip" pages in an adjacent frame according to markers embedded in the video. As shown in Figure 1, the video file in this presentation was indexed to a table of content, and that was done through markers embedded in the video file during the encoding process. These enabled students to click on items listed in the table of content (left window) in order to view its associated video along with its synchronized text and images in allocated areas within the presentation window.

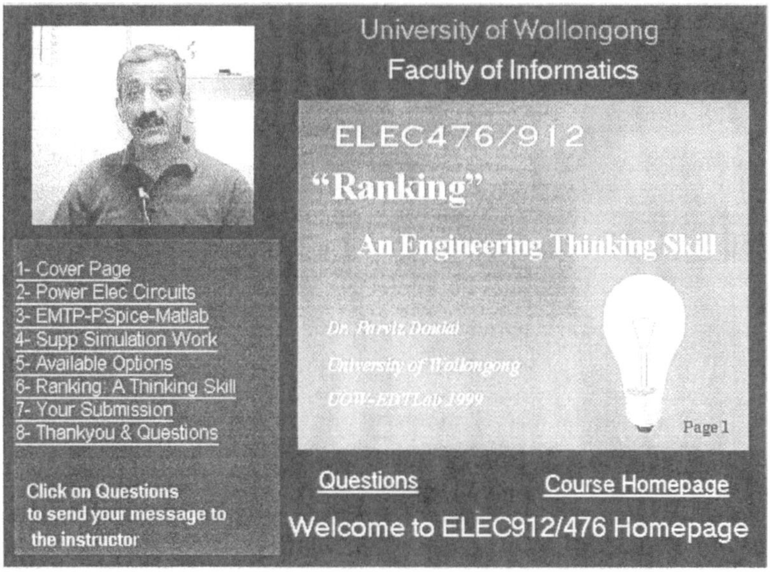

Figure 1. Stream video integration into a virtual classroom

An online questionnaire was administered to obtain information regarding student access to the subject homepage and its stream video integration in ELEC476/912 virtual learning environment. Survey results showed that students realized the benefits of technology-enhanced resources that were incorporated into their on-campus course delivery. Students' comments and feedback on the course content, the method of delivery and available tools and resources for this subject was archived in [14].

4.2 Case Study 2: ELEC101 Virtual Classroom

The Web Edition of "Fundamentals of Electrical Engineering (ELEC101)" is a simple virtual classroom environment that uses the real-time streaming technology to deliver synchronized Power Point slides (images) and audio files (the lecturer voice) over the Internet.

To ensure students using different computers of any power and different connections of any speed could retrieve the content of ELEC101 virtual classroom four options, namely plain, synchronized, controlled synchronized and power-point slide/script were provided. Figure 2 shows a screen caption of the cover page of ELEC101 World Wide Web Edition.

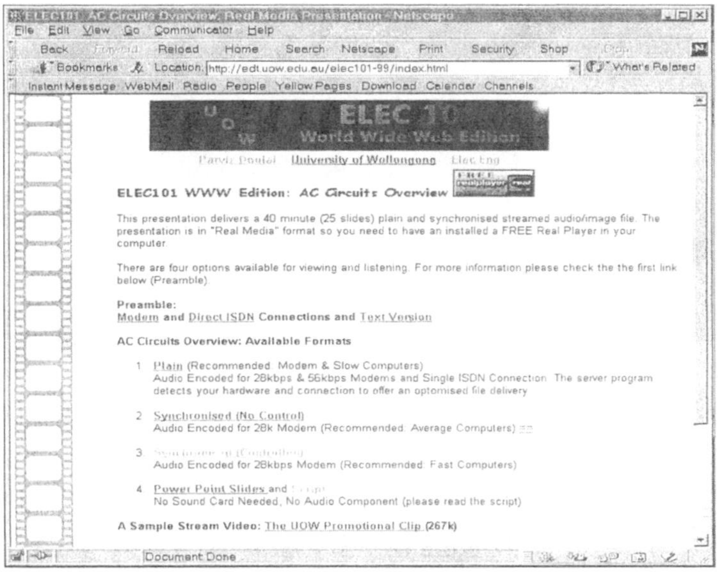

Figure 2. ELEC101 World Wide Web edition

Rather than replacing the conventional lecturing of ELEC101, the Web edition was designed and implemented to help students who need to review important pointers of major topics. Students need to have a freely available RealPlayer and perhaps a headphone set so that they can hear the lecture and view the overheads in computer laboratories or at home using a standard 56 kbps dial up connection on PC, Mac or UNIX platform.

In the plain format students first receive a page containing thumbnails of available overheads. The RealPlayer will start working as soon as students click on a thumbnail to view the actual overhead. Then, they step to the slide they are interested in, and hear the associated audio clip with each slide.

Students may control the RealPlayer operation, and they also have standard navigation tools. The RealPlayer may be used as a plug-in program or as a Netscape or Internet Explorer helper application. The latter means by clicking on the RealAudio icon, the browser lunches the player and from there, students control the player operation; recording, playback, rewind and so forth. They may also use standard previous and next buttons to move around. A screen caption of the plain format is shown in Figure 3.

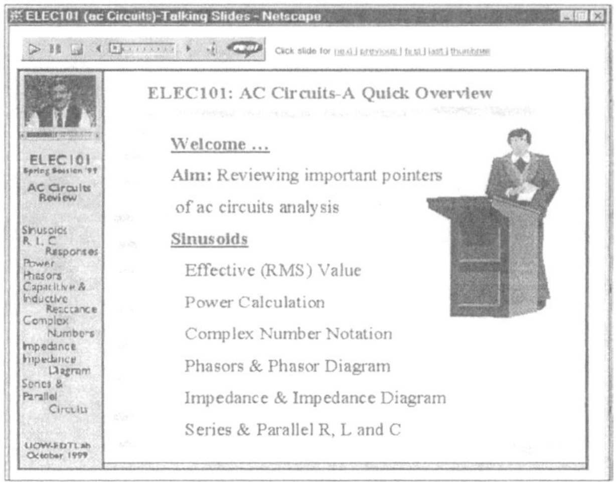

Figure 3. ELEC101: Plain mode of operation

In synchronized format student receive power point slides and their associated sound. The audio file automatically updates slides displayed as the lecture progresses. RealPlayer multiple controls were provided in this option. These include play, pause, volume-control and position-slider. Users can use the latter to move forward and backward through the presentation.

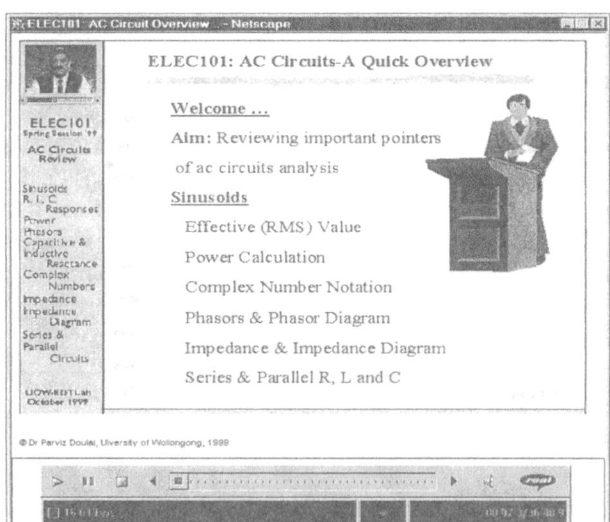

Figure 4. ELEC101: Synchronous mode of operation

The controlled synchronized option of the ELEC101 displays projected slides on the screen and plays the corresponding sound. In this mode of operation, students step to the slide they are interested in and start the player. While the audio is playing, it will automatically update the slide as the lecture progresses. Alternatively, students can jump to a new slide by clicking on thumbnails listed on the left frame, and the audio will jump to follow. To start listening to the audio from a particular slide, students may type the slide number in the space provided in control section and press the enter key. Figure 5 shows a screen caption of ELEC101 in a controlled synchronized mode of operation.

Provisions also were made for students using a computer without a sound card. In this case they view a slide on one window and read its corresponding text on another browser window.

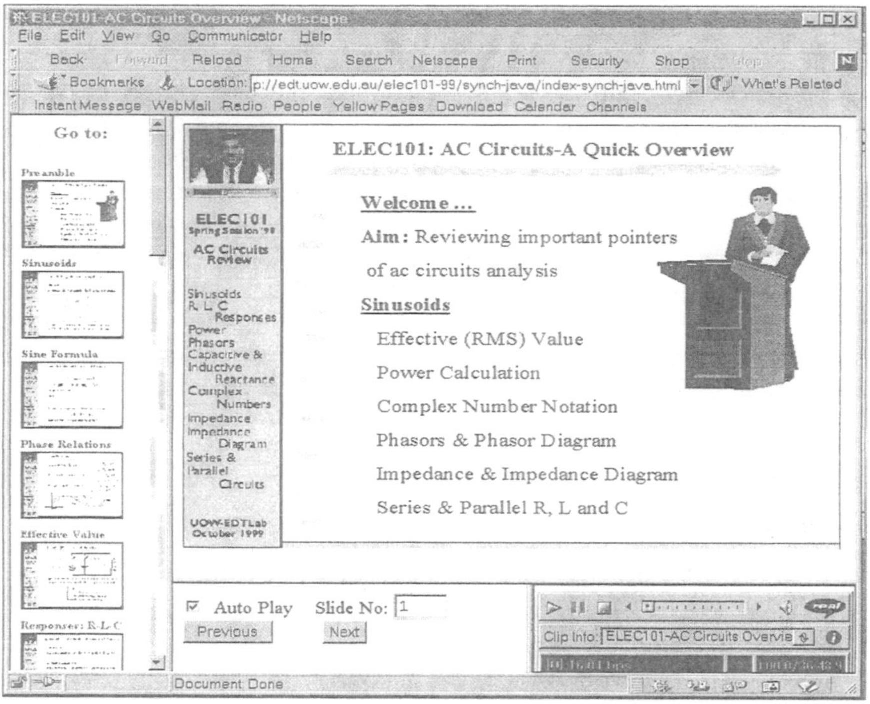

Figure 5. ELEC101: Controlled synchronous mode of operation

Implementation of the plain format is very simple provided the developer knows the technology and has some almost freely available tools. The "controlled synchronized" version of ELEC101 represents some challenges. This version uses JavaScript, Frames, and the RealAudio Plug-in.

Nowadays, the RealPlayer itself supports Java driven events. This basically means the development of synchronized audio and video files for network delivery is much easier, and can be done by almost everyone.

The ELEC101 virtual classroom environment was tested by a group of second year students using moderately high-speed connection (computer laboratories on campus) and low speed dial up connections (28.8kbps and higher modems). The setup performed with no interruptions or delay in delivering the subject content (sound and images). The entire concept of virtual classroom and the application of streamed and synchronized audio file were found by students very exciting and motivating. The setup is now available on Internet for public use [14].

5. CONCLUSION

The combination of powerful compression algorithms, extensive features that are associated with streaming servers and integration with the Web make it possible to use virtual learning environments effectively over narrow bandwidth networks. This paper explored the integration of the multimedia modules into a virtual learning environment. Real time streaming technology in an educational setting was examined and the process of applying streaming technology in education was briefly highlighted. Two examples of virtual learning environments using stream synchronized audio/video and image files were illustrated. It is envisaged that the usage of technology enabled methods in face-to-face university instruction results in a model that works equally well for distance students and learners in virtual campuses.

REFERENCE

[1] P. Doulai, "Preserving the quality of on-Campus education using resource-based approaches," *Proc. International WebCT Conference on Learning Technologies*, University of British Columbia, Vancouver, Canada, 1999, pp. 97-101.

[2] B. Hart-Davidson and R Grice, "Extending the dimensions of education: Designing, developing, and delivering effective distance-educ.," *Proc. of the IEEE Professional Communication Conference,* 2001, pp. 221-230.

[3] E. R. Ladd, J. R. Holt and H. A. Rumsey, "Washington state university's engineering management program distance education industry partnership," *Proc. of Portland International Conference on Management of Engineering and Technology*, 2001. pp. 302-306.

[4] P. Doulai, Smart and Flexible Campus: "Technology Enabled University Education," *Proc. of The World Internet and Electronic Cities Conference*, 2001, Iran, pp. 94-101.

[5] V. Trajkovic, D. Davcev etal, "Web-based virtual classroom," *Proc. of IEEE Conference on Technology of Object-Oriented Languages and Systems,* 2000, pp. 137-146

[6] W. Beuschel, "Virtual campus: scenarios, obstacles and experiences," *Proc. of IEEE Conference on System Sciences*, 1998, pp. 284-293.

[7] A. Zhang; Y. Song and M. Mieike, NetMedia: "Streaming multimedia presentations in distributed environments," *IEEE Multimedia*, Vol.9, 2002 pp. 56-73.

[8] P. Doulai, "Recent developments in Web-based educational technologies: A practical overview using in-house implementation," *Proc. of the International Power Engineering Conference,* 1999, Singapore, pp. 845-850.

[9] D. Fernandez, A. B. Garcia, D. Larrabeiti, A. Azcorra, P. Pacyna, and Z. Papir, "Multimedia services for distant work and education in an IP/ATM environment," *IEEE Multimedia*, Vol.8, 2001 pp. 68-77.

[10] Real Networks (Progressive Networks) http://www.real.com/

[11] Design and Management 1, "Introduction to Group Projects (ELEC195) Homepage," http://edt.uow.edu.au/elec195/welcome.ram

[12] S. Huang and H. Hu, "Integrating windows streaming media technologies into a virtual classroom environment," *Proc. of International Symposium on Multimedia Software Engineering*, 2000, pp. 411-418

[13] Apple QuickTime, http://www.apple.come/quicktime/

[14] The Educational Delivery Technology Laboratory (EDTLab), University of Wollongong, http://edt.uow.edu.au/edtlab/portfolio.html/

Chapter 2

WIDEBAND SPEECH AND AUDIO CODING IN THE PERCEPTUAL DOMAIN

L. Lin, E. Ambikairajah and W.H. Holmes
School of Electrical Engineering and Telecommunications, The University of New South Wales, UNSW Sydney 2052, Australia.

Abstract: A new critical band auditory filterbank with superior auditory masking properties is proposed and is applied to wideband speech and audio coding. The analysis and synthesis are performed in the perceptual domain using this filterbank. The outputs of the analysis filters are processed to obtain a series of pulse trains that represent neural firing. Simultaneous and temporal masking models are applied to reduce the number of pulses in order to achieve a compact time-frequency parameterization. The pulse amplitudes and positions are then coded using a run-length coding algorithm. The new speech and audio coder produces high quality coded speech and audio, with both temporal and spectral fidelity.

Key words: auditory filterbank, speech coding, simultaneous and temporal masking

1. INTRODUCTION

Current applications of speech and audio coding algorithms include cellular and personal communications, teleconferencing, secure communications etc. Historically, coding algorithms using incompatible compression techniques have been optimized for particular signal classes such as narrowband speech, wideband speech, high quality audio and high fidelity audio (CD quality). It is evident that a universal speech and audio coding paradigm is required to meet the diverse needs of the above applications. Low bit rate speech coders provide impressive performance above 4kbps for speech signals. But do not perform well on music signals. Similarly, transform coders perform well for music signals, but not for speech signals at lower bit rates.

Speech and general audio coders are usually quite different – for speech one of the main tools is a model of the speech production process, whereas for audio more attention is paid to modeling the human auditory system, since a source model is usually not feasible. The new MPEG-4 standard for multimedia communication includes a scalable audio codec supporting transmission at bit rates from 2 to 64kbps. However, in order to achieve the highest audio quality with the full range of bit rates, MPEG-4 actually employs three types of codec. For lower bit rates, a parametric codec (Harmonic Vector Excitation Coding) is used which encodes at 2-4kbps for speech with an 8kHz sampling frequency, and at 4-16kbps for speech and audio with 8 or 16kHz sampling frequency. A Code Excited Linear Predictive (CELP) codec is used for the medium rate – i.e. 6-24kbps at 8 or 16kHz sampling frequency. Time-frequency (TF) codecs, including the MPEG-2 AAC and Twin VQ codecs are used for the higher bit rates, requiring 16-64kbps at a sampling frequency of 8kHz.

There is therefore a need for high quality coders that can work equally well with either speech or general audio signals. In this work we propose a scheme for a universal coder that can handle both wideband speech and audio signals. This coder is based on a new auditory filterbank model, and is a further development of the speech and audio coding scheme initially proposed by Ambikairajah *et al.* [3], in which the analysis and synthesis of the speech and audio signals take place in the perceptual domain.

1.1 Coding using Auditory Filterbanks

In recent years parallel auditory filterbanks such as the Gammatone filterbank [5,13] have outperformed the conventional transmission line auditory model [1,12] in terms of computational simplicity. They have applications in various types of signal processing required to model human auditory filtering. Gammatone auditory filters were first proposed by Flanagan [5] to model basilar membrane motion, and were subsequently used by Patterson *et al.* [13] as a reasonably accurate alternative for auditory filtering. They have since become very popular. Robert and Eriksson [15] applied them to produce a nonlinear active model of the auditory periphery, and Kubin and Kleijn [7] applied them to speech coding.

In the wideband speech and audio coder proposed by Ambikairajah *et al.* [3], the analysis is performed in the auditory domain by using Gammatone filters to obtain an auditory-based time-frequency parameterization of the input signal in the form of critical band pulse trains. This parameterization approximates the patterns of neural firing generated by the auditory nerves, and preserves the temporal information present in speech and music. An advantage of this parameterization is its ability to scale easily between different sampling rates, bit rates and signal types.

Adequate modeling of the principal behavior of the peripheral auditory systems is still a difficult problem. An important shortcoming of Gammatone filters is that they do not provide an accurate frequency domain description of the tuning curves because of their flat upper-frequency slopes. In this work we propose a new parallel auditory filterbank based on the critical band scale. The filterbank models psychoacoustic tuning curves obtained from the well-known masking curves [16,17]. The new auditory filters, which have a steeper upper-frequency slope, achieve high frequency domain accuracy and are computationally efficient. The new filterbank is then applied to wideband speech and audio coding under the same paradigm as in [3]. Auditory masking is applied to eliminate redundant information in the critical band pulse trains. A technique to code the pulse positions and amplitudes based on a run-length coding algorithm is also proposed.

This chapter is organized as follows: Section 2 presents the design techniques for the new critical band auditory filterbank. Section 3 describes the auditory-filterbank-based speech and audio coding scheme, including the reduction of redundancy in the pulse trains and the quantization and coding techniques for the pulse amplitudes and positions.

2. DESIGN OF A CRITICAL BAND AUDITORY FILTERBANK

A filterbank that models the characteristics of the human hearing system will have many desirable features and can have wide applications in speech and audio processing. It is very difficult and costly to experimentally observe the motion of the basilar membrane in a fully functional cochlea. We present here an inexpensive method for generating psychoacoustic tuning curves from the well-known auditory masking curves [16,17]. Then two approaches to obtain the critical band filterbank that model these tuning curves are introduced. The first approach is based on the Log-Modeling technique for filter design, which gives very accurate results. The second approach uses a unified transfer function to represent each filter in the critical band filterbank.

2.1 Generation of Psychoacoustic Tuning Curves from Masking Curves

Masking is usually described as the sound-pressure level of a test sound necessary to be barely audible in the presence of a masker. Using narrow-band noise of a given center frequency and bandwidth as maskers and a pure tone as the test sound, masking patterns have been obtained by Zwicker and Fastl [16,17]. The effect of masking produced by narrow-band maskers is level dependent. The five curves plotted as solid lines in Fig. 1 are the masking patterns centered at 1 kHz at the five different levels L_G = 20, 40, 60, 80 and 100 dB [17].

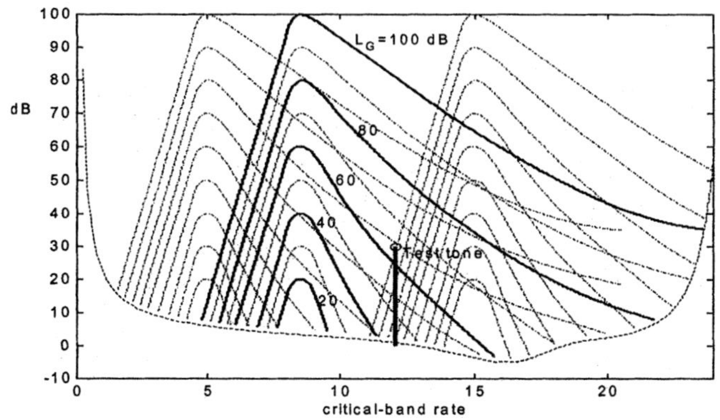

Figure 1. Masking curves obtained by interpolation and shifting

It is known that the shapes of the masking patterns for different center frequencies and different levels are very similar when plotted using the critical band rate scale. Hence masking curves at different center frequencies can be obtained by simply shifting the available masking curves at f_c = 1 kHz. Masking curves at levels other than L_G = 20, 40, 60, 80 and 100 dB can be generated through interpolation. The masking curves obtained through interpolation and shifting are shown in Fig. 1 by the dashed lines.

The tuning curves can be obtained from the masking curves as follows. The first step is to fix a test tone at a particular frequency and level. Then the masking curves with different center frequencies that are just able to mask the testing tone are found and the corresponding levels are noted. Plotting the levels as a function of the center frequencies provides the tuning curve at that test tone frequency (Fig. 2).

The magnitude response of the basilar membrane (or auditory filters) can be obtained by vertically reversing and scaling the tuning curves in Fig. 2. This is shown in later subsections in Fig. 3 and 4 by the dashed lines. More details can be found in [11]. The tuning curves are consistent with the measurement of nerve tuning curves [8] and the basilar membrane response [14]. Two auditory filter design techniques that model the magnitude response accurately are introduced in the next subsection.

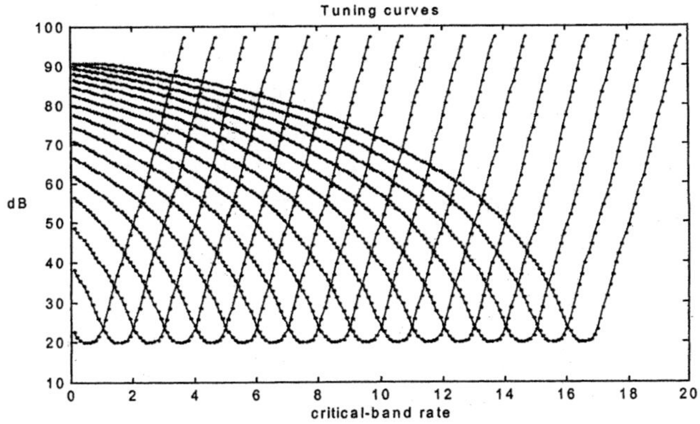

Figure 2. Psychoacoustic tuning curves

2.2 Filterbank Design by the Log-Magnitude Modeling Technique

It is well known that the human auditory system gives rise to a perception of loudness that closely follows a logarithmic scale. Log-magnitude modeling is a technique for IIR digital filter design [6]. This technique has also been applied in [10] to the modeling of auditory tuning curves. The result is a very accurate model that matches the magnitudes of the tuning curves. The criterion for auditory filter design is based on the minimization of the difference between the log-magnitude of the desired basilar membrane frequency response and a pole-zero filter. The transfer function of one filter in a critical band rate filterbank can be written as

$$G(z)=\frac{B(z)}{A(z)}=\frac{\sum_{i=0}^{Q} b_i z^{-i}}{\sum_{i=0}^{P} a_i z^{-i}}, \qquad (1)$$

where a_i and b_i are the filter parameters, P is the number of poles, and Q is the number of zeroes. The filter design technique minimize the sum of squared differences, on a logarithmic scale, between a given set of spectral amplitudes $D(\omega_k)$ and the magnitude response of $G(e^{j\omega_k})$ sampled at the same frequencies:

$$J=\sum_{k=0}^{N-1}\left[\log D(\omega_k)-\log\left|G(e^{j\omega_k})\right|\right]^2, \qquad (2)$$

where $\omega_k=2\pi k/N$, $k = 0, 1, \ldots, N$-1, is a set of uniformly spaced frequencies and $D(\omega_k)$ is the desired basilar membrane frequency response (positive magnitude values) at a certain center frequency.

The minimization of J with respect to the parameters a_i and b_i is a nonlinear problem. To avoid gradient-based optimization, an iterative procedure originally proposed in [6] is used. The minimization index at the mth step can be written as

$$J^m = \sum_{k=0}^{N-1} W^m(\omega_k)\left|D(\omega_k)A^m(e^{j\omega_k}) - B^m(e^{j\omega_k})\right|^2$$
$$= \begin{bmatrix} \mathbf{a}^{m^T} & \mathbf{b}^{m^T} \end{bmatrix} \begin{bmatrix} \mathbf{E} & -\mathbf{C} \\ -\mathbf{C}^T & \mathbf{F} \end{bmatrix} \begin{bmatrix} \mathbf{a}^m \\ \mathbf{b}^m \end{bmatrix} \quad (3)$$

The filter at step m is computed from

$$\begin{bmatrix} \mathbf{E} & -\mathbf{C} \\ -\mathbf{C}^T & \mathbf{F} \end{bmatrix} \begin{bmatrix} \mathbf{a}^m \\ \mathbf{b}^m \end{bmatrix} = \begin{bmatrix} J^m_{\min} \\ \mathbf{0} \end{bmatrix}, \quad (4)$$

where

$$\mathbf{a}^m = \begin{bmatrix} 1 & a_1^m & a_2^m & \cdots & a_P^m \end{bmatrix}^T, \ \mathbf{b}^m = \begin{bmatrix} b_0^m & b_1^m & b_2^m & \cdots & b_Q^m \end{bmatrix}^T.$$

The solution of (4) is used to update the weight function in (3) and the process is then repeated. The complete algorithm converges to a sufficiently small error J^m within 2 to 3 iterations. The details of this procedure can be found in [6,10]. A critical band filterbank of 17 filters covering the frequency range of 50 Hz to 4000 Hz was obtained by this design technique. The frequency response of the 17 filters is shown in Fig. 3 by the solid lines, together with the vertically flipped tuning curves by the dashed lines. These filters are minimum-phase IIR filters with 8 poles and 7 zeros. The magnitude responses of the digital filters are almost indistinguishable from the true tuning curves.

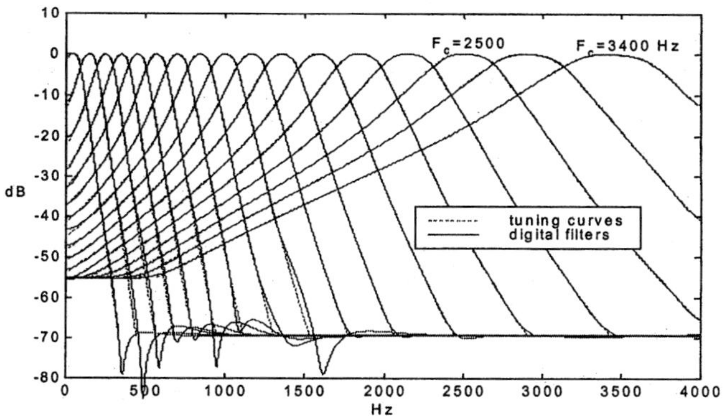

Figure 3. Digital filters obtained by Log-Magnitude modeling (solid lines: digital filters; dashed lines: tuning curves)

2.3 Filterbank Design by Direct Modeling Approach

A unified digital filter model is proposed in [11] to represent the frequency characteristics of all the tuning curves. The transfer function of one auditory filter in a critical band filterbank is expressed in the z-domain by

$$G(z) = \frac{(1-r_0 z^{-1})\,(1-2r_B \cos(2\pi f_B / f_s) z^{-1} + r_B^2 z^{-2})}{(1-2r_A \cos(2\pi f_A / f_s) z^{-1} + r_A^2 z^{-2})^4} \, . \tag{5}$$

The parameters in (5) are given by

$$f_A = \sqrt{f_c^2 + B_w^2}, \qquad r_A = e^{-2\pi B_w / f_s}\,, \tag{6}$$

where f_s is the sampling frequency. The critical bandwidth B_w and the central frequency f_c in (6) are calculated from the following equations [16, 17]:

$$\begin{aligned} Z_c &= 13\tan^{-1}(0.76 f_c / 1000) + 3.5\tan^{-1}(f_c / 7500)^2 \\ B_w &= 25 + 75[1 + 1.4(f_c / 1000)^2]^{0.69} \end{aligned} \tag{7}$$

where Z_c is the critical band rate in Bark corresponding to f_c. The spacing of Z_c is linear on a critical band scale.

The parameter r_0 is chosen as $r_0 = 0.955$. The term $(1-2r_B \cos(2\pi f_B / f_s) z^{-1} + r_B^2 z^{-2})$ produces a notch filter with a sharp dip at a point to the right of the center frequency f_c so that the upper-frequency slope of the overall filter is steep enough. The parameter r_R is chosen as $r_R = 0.985$. To ensure the notch happens at a frequency location about 60 dB lower than the center frequency f_c, the empirical formula that we obtained can be used to choose f_B:

$$f_B = 117.5(f_c/1000)^2 + 1135.5(f_c/1000) + 277.0 \ , \tag{8}$$

where f_c is in Hz.

The frequency responses of five filters at critical bands 4, 7, 10, 13 and 16 are plotted in Fig. 4, together with the corresponding tuning curves. The modeling accuracy of this direct modeling approach is acceptable and is more straightforward than the log-magnitude modeling approach.

Our filters are also compared with the well-known Gammatone auditory filters [5,13]. Our filters have steeper upper-frequency slopes, which is desirable for both accurate modeling of the masking effect and noise suppression. Critical band filters designed using this method can achieve both high frequency domain accuracy and computational efficiency. Next we will apply the critical band auditory filterbank to speech and audio processing.

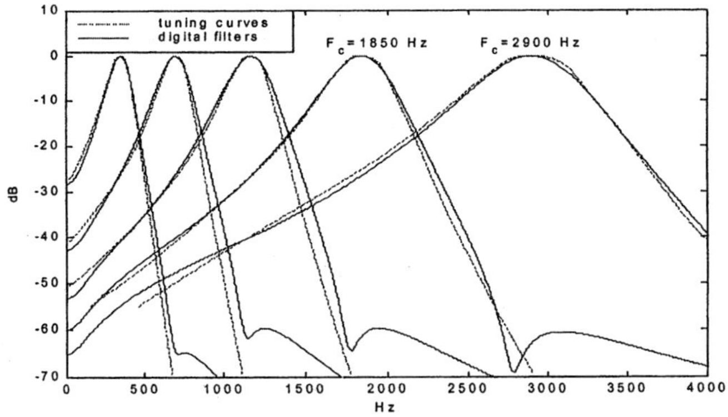

Figure 4. Auditory filters obtained by direct modeling (solid line: digital filters; dashed: tuning curves)

3. PERCEPTUAL DOMAIN BASED SPEECH AND AUDIO CODING

3.1 Speech/audio Coding Using an Auditory Filterbank

The speech and audio coding system implemented in this work is an IIR/FIR analysis/synthesis scheme as described in [9] and also shown in Figs. 5 and 6. Other possible analysis/synthesis filterbank implementations can also be found in [9].

Each IIR analysis filter has 8 poles and 3 zeros. The analysis filterbank can also be implemented in FIR form [3,7], but at least 100 coefficients are required for each FIR filter to approximate the impulse response of the IIR filter with reasonable accuracy. The auditory filterbank is also approximately power-complementary. That is,

$$\sum_{i=1}^{M} | G_i(e^{j\omega}) |^2 \approx C \ , \tag{9}$$

where $G_i(e^{j\omega})$ is the frequency response of the analysis filter at the ith channel and M is the total number of channels. If we choose the synthesis filters as

$$h_i(n) = g_i(-n) \ \text{for} \ i = 1 \ ... \ M \, , \tag{10}$$

then the synthesis filterbank is implemented using FIR filters obtained by time-reversal of the impulse responses of the corresponding analysis filters. The reconstruction is nearly perfect – i.e.

$$\sum_{i=1}^{M} g_i(n) * h_i(n) \approx C\delta(n) \, . \tag{11}$$

Each FIR synthesis filter has 128 coefficients, so that an 8 ms delay is required to make the filter causal if f_s =16 kHz.

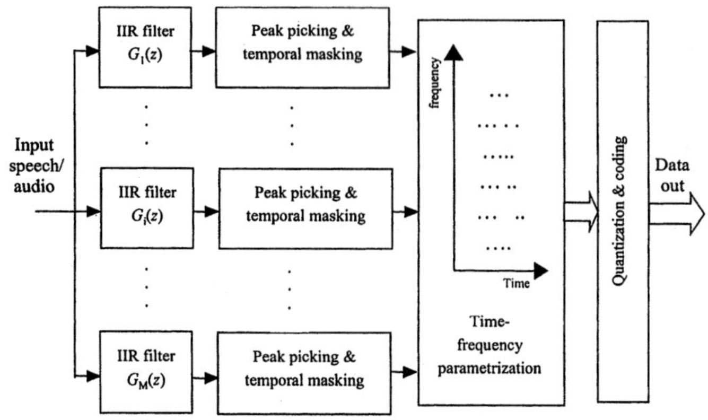

Figure 5. Speech/audio using an auditory filterbank: Analysis system.

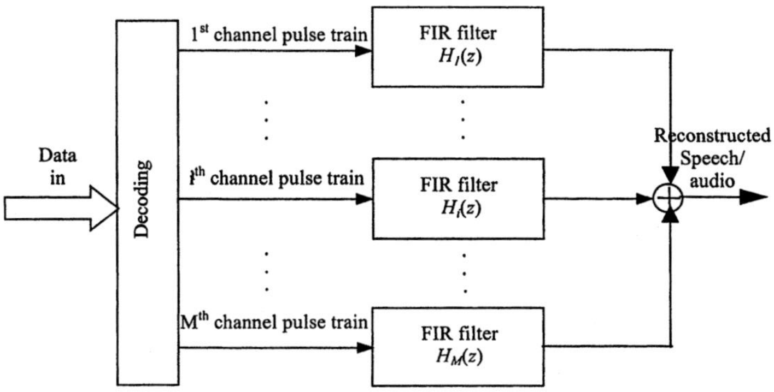

Figure 6. Speech/audio using an auditory filterbank: Synthesis system.

3.2 Auditory Masking

The output of each filter is half-wave rectified, and the positive peaks of the critical band signals are located. Physically, the half-wave rectification process corresponds to the action of the inner hair cells, which respond to movement of the basilar membrane in one direction only. Peaks correspond to higher rates of neural firing at larger displacements of the inner hair cell from its position at rest. This process results in a series of critical band pulse trains, where the pulses retain the amplitudes of the critical band signals from which they were derived.

In recognition of the fact that lower power components of the critical band signals are rendered inaudible by the presence of larger power components in neighboring critical bands, a simultaneous masking model is employed. Weak signal components become inaudible by the presence of stronger signal components in the same critical band that precede or follow them in time, and this is called temporal masking. When the signal precedes the masker in time, it is called pre-masking; when the signal follows the masker in time, the condition is called post-masking. A strong signal can mask a weaker signal that occurs after it and a weaker signal that occurs before it [2,16,17]. Both temporal pre-masking and temporal post-masking are employed in this work to reduce the number of pulses.

3.2.1 Simultaneous Masking

In the implementation described a simultaneous masking model similar to that used in MPEG [4] was employed to calculate the masking threshold L_i (dB) for the ith critical band, however the optimum simultaneous masking model for this scheme has yet to be determined. The simultaneous masked pulse train $x_i'(n)$ for the ith critical band was obtained from pulses in the unmasked pulse train $x_i(n)$ whose amplitudes were below the masking threshold calculated for each critical band were considered inaudible, and were set to zero

$$x_i'(n) = \begin{cases} x_i(n), & x_i(n) \geq 10^{L_i/20} \\ 0, & x_i(n) < 10^{L_i/20} \end{cases}, \qquad (12)$$

Note that for each 32 ms frame, the gain of each critical band is calculated based only on the non-zero pulse amplitudes. The purpose of applying simultaneous masking is to produce a more efficient and perceptually accurate parameterization of the firing pulses occurring in each band. Experiments revealed that simultaneous masking removed an average of around 10% of the pulses without altering the quality of the reconstructed speech in any way.

3.2.2 Temporal Post-masking

The masking threshold $y_i(n)$ for temporal post-masking decays approximately exponentially following each pulse, or neural firing. A simple approximation to this masking threshold, introduced in [3], is

$$y_i(n) = \begin{cases} x_i'(n), & x_i'(n) > c_0\, y_i(n-1) \\ c_0\, y_i(n-1), & \textit{otherwise} \end{cases}, \qquad (13)$$

where $x_i'(n)$ is the ith of M = 21 simultaneous masked critical band pulse train signals, $c_0 = \exp(-\tau_i)$, and n is the discrete time sample index. The time constants τ_i, $1 \le i \le M$, were determined empirically by listening to the quality of the reconstructed speech, and values between 0.008 ($i = 1$) and 0.03 ($i = 21$) were chosen. All pulses with amplitudes less than the masking threshold $y_i(n)$ were discarded. The thresholds are shown in Fig. 7 by the dashed line, where the filled spikes are the pulses to be kept after applying post-masking.

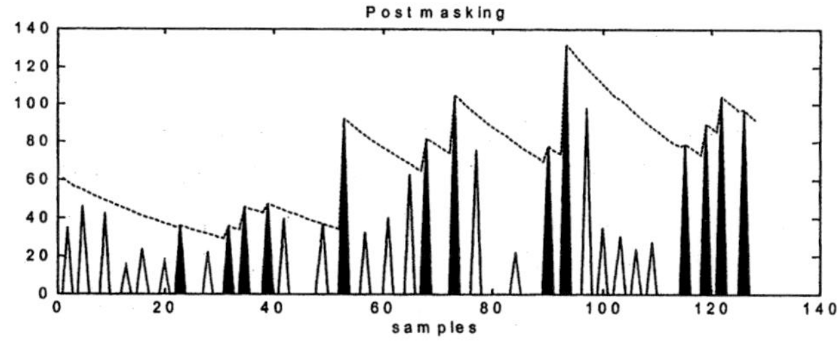

Figure 7. Pulse reduction using post-masking (solid lines: pulses; dashed lines: thresholds)

3.2.3 Temporal Pre-masking

Pre-masking is also allowed for in this work. The masking threshold $z_i(n)$ for this temporal pre-masking is chosen as

$$z_i(n-1) = \begin{cases} x_i''(n-1), & x_i''(n-1) > c_1 z_i(n) \\ c_1 z_i(n), & otherwise \end{cases} \quad (14)$$

where $x_i''(n)$ is the ith critical band pulse train after post-masking, and c_1 is chosen as $c_1 = \exp(-3\tau_i)$ to simulate the fast exponential decay of pre-masking. All pulses with amplitude less than the masking threshold $z_i(n)$ were discarded. This is shown in Fig. 8, where the filled spikes are the pulses to be kept after applying pre-masking. A reduction rate of 10% can be achieved by pre-masking on the pulses obtained after post-masking.

The purpose of applying masking is to produce a more efficient and perceptually accurate parameterization of the firing pulses occurring in each band. Experiments show that the application of temporal masking reduces the overall pulse number to about 0.70N (where N is the frame size) while maintaining transparent quality of the coded speech and audio. This is a significant improvement over the pulse number of 1.26N in the previous application [3], which used Gammatone filters in the front end. The improvement is mainly due to the spectral shape of the new auditory filters used in this work.

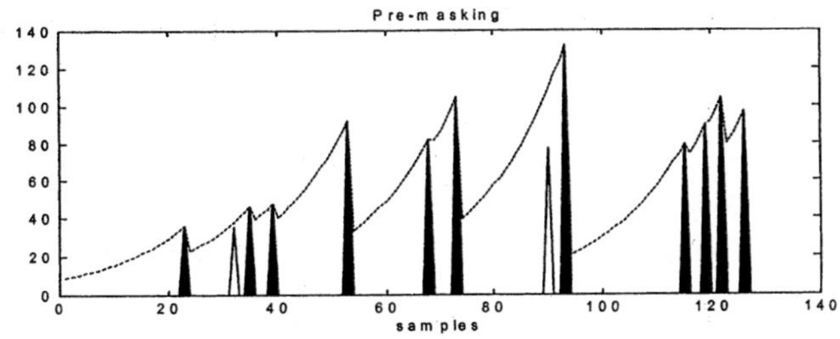

Figure 8. Pulse reduction using pre-masking (solid lines: pulses; dashed lines: thresholds.)

3.2.4 Thresholding

The pulses in the silent frames obtained after auditory filtering and peak picking are most likely due to background and quantization noise. These pulses are at random positions and their magnitudes are very small, so that the sound synthesized from these pulses are inaudible. By thresholding, these pulses can be eliminated without affecting the quality of the synthesized signal. A simple approach is to choose the threshold based on the silent frames at the beginning of the coding process.

3.3 Quantization and Coding

The pulse train in each critical band after redundancy reduction was finally normalized by the mean of its non-zero pulse amplitudes across the frame. Thus, the parameterization consists of the critical band gains (incorporating the normalization factors) and a series of critical band pulse trains with normalized amplitudes. For each frame, the signal parameters requiring for coding are the gains of the critical bands and the amplitudes and positions of the pulses.

3.3.1 Pulse Amplitudes

Each critical band gain is quantized to 6 bits and the amplitude of each pulse is quantized to 1 bit, which does not result in any perceivable deterioration in the quality of the reconstructed speech or audio signal. Alternatively, vector quantization can be adapted to reduce the bits required for coding the amplitude [3].

3.3.2 Pulse Positions

The pulse positions are coded using a new run-length coding technique. After temporal masking and thresholding, most locations on the time-frequency map have zero pulses. This suggests that we can just code the relative positions of neighboring pulses or the numbers of zeros between them. Specifically, the data in all channels with one frame is concatenated into one large vector and is scanned for pulses. Then the number of zeros preceding each pulse is coded using 7 bits. An example is shown below

$$\ldots 0\,0\,1.2\underbrace{0\,0\ldots0\,0}_{\substack{128\text{ zeros}\\(0000000)}}\underbrace{0\,0\ldots0\,0}_{\substack{120\text{ zeros}\\(1111000)}}1.5\underbrace{0\,0\ldots0\,0}_{\substack{89\text{ zeros}\\(1011001)}}0.8\,0\,0\ldots$$

If the number of zeros is over 128, a code word of 0000000 is generated and the counting of zeros restarts after the 128 zeros. If during the decoding process, seven consecutive zeros are encountered, then no pulse will be generated and the decoding carries on to the next code word. This coding strategy is a form of run-length coding and is lossless.

The overall average bit rate resulting from this coding scheme is 58 kbps. This is an improvement upon the 69.7 kbps in the previous work [3]. By exploring the statistical correlations and redundancy among the pulses, Huffman or arithmetic coding can be applied to further reduce the bit rate.

The synthesis process starts with decoding to obtain the pulse train for each channel, and then filtering the pulse train by the corresponding FIR synthesis filter. Summing the outputs from all filters results in the reconstructed speech or audio signal, which is perceptually the same as the original. The results at different stages are shown in Figs. 9-12, where Fig. 9 is the original speech signal, Fig. 10 shows the pulses obtained from peak-picking, Fig. 11 shows the pulses retained after applying auditory masking, and Fig. 12 is the reconstructed speech.

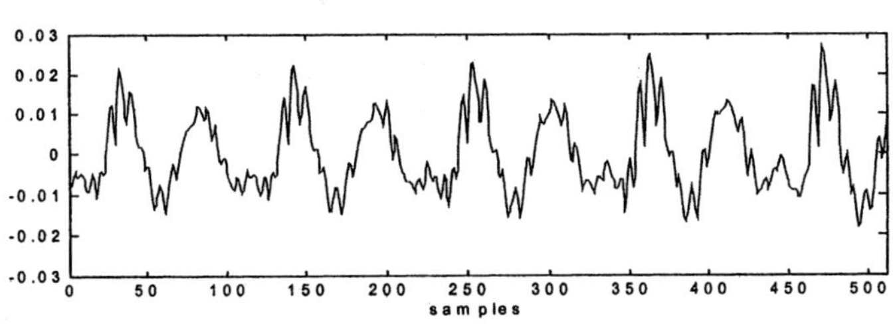

Figure 9. Original speech

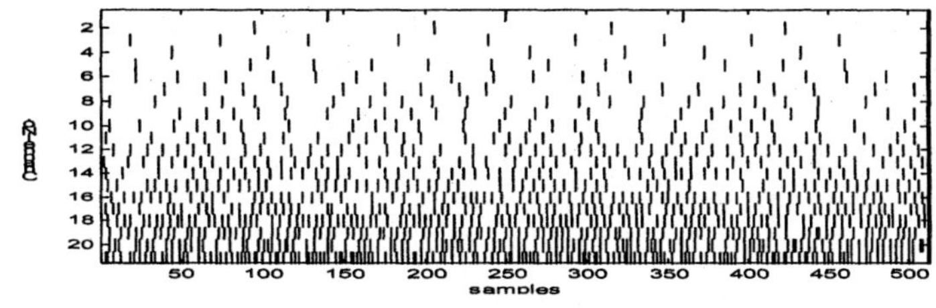

Figure 10. Pulses obtained from peak-picking

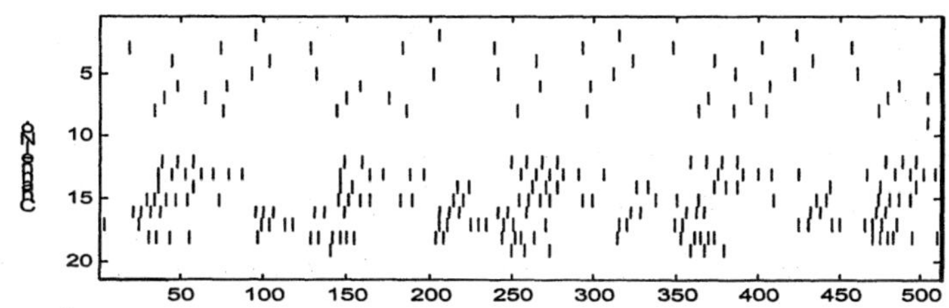

Figure 11. Pulses retained after auditory masking

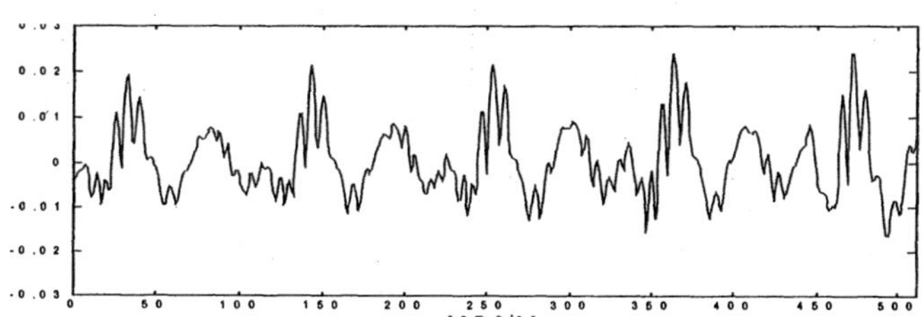

Figure 12. Reconstructed speech

4. CONCLUSIONS

Design techniques for a new critical band auditory filterbank that models the psychoacoustic tuning curves have been proposed. The auditory filterbank has been applied to speech and audio coding. The filterbank is implemented as an IIR/FIR analysis/synthesis scheme to reduce computation. Auditory masking is applied to reduce the number of pulses. A simple run-length coding algorithm is used to code the positions of the pulses. The reconstructed speech or audio signals are perceptually transparent. The overall average bit rate resulting from this coding scheme is 58kbps. The filterbank has superior masking properties and the auditory-system-based coding paradigm produces high quality coded speech or audio, is highly scalable, and is of moderate complexity. Current research involves investigation into to the use of Huffman coding or arithmetic coding techniques to further reduce the bit rate by examining the statistical correlation and redundancy among the pulses.

REFERENCES

[1] Ambikairajah, E., Black, N.D. and Linggard, R., "Digital filter simulation of the basilar membrane", *Computer Speech and Language*, 1989, vol. 3, pp. 105-118.

[2] Ambikairajah, E., Davis, A.G., and Wong, W.T.K., "Auditory masking and MPEG-1 audio compression", *Electr. & Commun. Eng. Journal*, vol. 9, no. 4, August 1997, pp. 165-197.

[3] Ambikairajah, E., Epps, J. and Lin, L., "Wideband speech and audio coding using Gammatone filter banks", *Proc. ICASSP*, 2001, pp. 773-776.

[4] Black, M. and Zeytinoglu, M., "Computationally efficient wavelet packet coding of wide-band stereo audio signals ", *Proc. ICASSP*, 1995, pp. 3075-3078.

[5] Flanagan, J.L., "Models for approximating basilar membrane displacement", *Bell Sys. Tech. J*, 1960, vol. 39, pp. 1163-1191.

[6] Kobayashi, T. and Imai, A., "Design of IIR digital filter with arbitrary log magnitude function by WLS techniques", *IEEE Trans. ASSP*, vol. ASSP-38, 1990, pp. 247-252.

[7] Kubin, G. and Kleijn, W.B., "On speech coding in a perceptual domain", *Proc. ICASSP*, 1999, pp. 205-208.

[8] Liberman, M.C. "Auditory-nerve response from cats raised in a low-noise chamber", *J. Acoust. Soc. Am.*, vol. 63, 1978, pp. 442-455.

[9] Lin, L., Holmes, W.H. and Ambikairajah, E., "Auditory filter bank inversion", *Proc. ISCAS 2001*, 2001. Vol. 2 pp: 537 –540.

[10] Lin, L., Ambikairajah, E. and Holmes, W.H., "Log-magnitude modelling of auditory tuning curves", *Proc. ICASSP*, 2001, pp. 3293-3296.

[11] Lin, L., Ambikairajah, E. and Holmes, W.H., "Auditory filterbank design using masking curves", *Proc. EUROSPEECH 2001*, pp. 411-414.

[12] Lyon, R.F., "A computational model of filtering detection and compression in the cochlea", *Proc. ICASSP*, 1982, pp. 1282-1285.

[13] Patterson, R.D., Allerhand, M., and Giguere, C., "Time-domain modelling of peripheral auditory processing: a modular architecture and a software platform", *J. Acoust. Soc. Am.*, vol. 98, 1995, pp. 1890-1894.

[14] Rhode, W.S., "Observation of the vibration of the basilar membrane of the squirrel monkey using the Mossbauer technique", *J. Acoust. Soc. Am.*, vol. 49, 1971, pp. 1218-1231.

[15] Robert, A. and Eriksson, J., "A composite model of the auditory periphery for simulating responses to complex sounds", *J. Acoust. Soc. Am.*, vol. 106, 1999, pp. 1852-1864.

[16] Zwicker, E. and Zwicker, U.T., "Audio engineering and psychoacoustics: matching signals to the final receiver, the human auditory system", *J. Audio Eng. Soc.*, vol. 39, No. 3, 1991, pp. 115-125.

[17] Zwicker, E. and Fastl, H., *Psychoacoustics: Facts and models*. Springer-Verlag, 1999.

Chapter 3

RECOGNITION OF ENVIRONMENTAL SOUNDS USING SPEECH RECOGNITION TECHNIQUES

Michael Cowling and Renate Sitte.
Griffith University, Gold Coast, Qld 9726,, Australia

Abstract: This paper discusses the use of speech recognition techniques in non-speech sound recognition. It analyses the different techniques used for speech recognition and identifies those that can be used for non-speech sound recognition. It then performs benchmarks on these techniques and determines which technique is better suited for non-speech sound recognition. As a comparison, it also gives results for the use of learning vector quantization (LVQ) and artificial neural network (ANN) techniques in speech recognition.

Key words: non-speech sound recognition, environmental sound recognition, artificial neural networks, learning vector quantization, dynamic time warping, long-term statistics, mel-frequency cepstral coefficients, homomorphic cepstral coefficients

1. INTRODUCTION

It has long been a goal of researchers around the world to build a computer that displays features and characteristics similar to those of human beings. The research of Brooks [1] is an example of developing human-like movement in robots. However, another subset of this research is to develop machines that have the same sensory perception as human beings. This work finds its practical application in the wearable computer domain (e.g. certain cases of deafness where a bionic ear (cochlea implant) cannot be used.)

Humans use a variety of different senses in order to gather information about the world around them. If we were to list the classic five human senses in order of importance, it is generally accepted that we would come up with the sequence: vision, hearing, touch, smell, taste.

Vision is undoubtedly the most important sense with hearing being the next important and so on. However, despite the fact that hearing is a human beings second most important sense, it is all but ignored when trying to build a computer that has human like senses. The research that has been done into computer hearing revolves around the recognition of speech, with little research done into the recognition of non-speech environmental sounds.

This chapter expands upon the research done by the authors [2, 3]. In these papers, a prototype system is described that recognizes 12 different environmental sounds (as well as performing direction detection in 7 directions in a 180° radius). This system was implemented using Learning Vector Quantization (LVQ), because LVQ is able to produce and modify its classification vectors so that multiple sounds of a similar nature are still considered as separate classes. However, no comparative testing was done to ensure that LVQ was the best method for the implementation of a non-speech sound classification system.

Therefore, this chapter will review the various techniques that can be used for non-speech recognition and perform benchmark tests to determine the technique most suited for non-speech sound recognition. Due to lack of research into non-speech classification systems, this chapter will focus on using speech and speaker recognition techniques applied to the domain of environmental non-speech sounds.

The remainder of this chapter will be split into four sections. The first section will discuss techniques that have been previously used for speech recognition and identify those techniques that could also be applied to non-speech recognition. The second section will show the results of benchmarks on these techniques and also compare their performance with results for speech recognition. The third section of this chapter will discuss these results. Finally, the fourth section will conclude and suggest areas for future research.

2. SELECTION OF TECHNIQUES

Research into speech recognition began by reviewing the literature and finding techniques that had previously been used for speech/speaker recognition. Techniques for both feature extraction and system learning were analyzed and those techniques that could be used for non-speech sound recognition were identified. These techniques were then benchmarked and results will be presented in the Results section.

2.1 Feature Extraction

For feature extraction, the literature review showed that speech recognition relies on only a few different types of feature extraction techniques (each with several different variations). Eight techniques were selected as possible candidates for feature extraction of non-speech sounds. These were:

- Frequency Extraction
- LPC Cepstral Coefficients
- Homomorphic Cepstral Coefficients
- Mel Frequency Cepstral Coefficients
- Mel Frequency LPC Cepstral Coefficients
- Bark Frequency Cepstral Coefficients
- Bark Frequency LPC Cepstral Coefficients
- Perceptual Linear Prediction Features

In addition, it was found that emerging research in speech recognition suggests the use of time-frequency techniques such as wavelets. Due to the emerging nature of this research, these techniques will not be included in this comparison. However, for an insight into how wavelets can be used for speaker recognition, please refer to the chapter in this volume by Michael Orr et al, "A Novel Dual Adaptive Approach to Speech Processing".

A specific investigation was then performed for each of these eight techniques. This investigation revealed that techniques based on LPC Cepstral Coefficients were based on the idea of a *vocoder*, which is a simulation of the human vocal tract. Since the human vocal tract does not produce environmental sounds, these techniques are not appropriate for recognition of non-speech sounds.

In addition, Lilly [4] mentions that the results of the Mel Frequency Based Filter and the Bark Frequency filter are similar, mainly due to the similar nature of these filters. Gold [5] also mentions that PLP and Mel Frequency are similar techniques. Based on these previous findings, only the more popular Mel Frequency technique was selected for benchmarking.

This leaves three feature extraction techniques to be tested:

- Frequency Extraction
- Homomorphic Cepstral Coefficients
- Mel Frequency Cepstral Coefficients

2.2 System Learning

The following system learning techniques are commonly used for speech/speaker recognition or have, in the past, been used for this application domain. They are:

- Dynamic Time Warping (DTW)
- Hidden Markov Models (HMM)
- Vector Quantization (VQ) / Learning Vector Quantization (LVQ)
- Self-Organizing Maps (SOM)
- Ergodic-HMM's
- Artificial Neural Networks (ANN)
- Long-Term Statistics

To aid in selection of techniques, comparison tables were built (using [5, 6, 7, 8]) to compare the different feature extraction and classification methods used by each of these techniques.

The comparison tables showed that some of these techniques, by their very nature, could not be used for non-speech sound recognition. Any of the techniques that use subword features are not suitable for non-speech sound identification. This is because environmental sounds lack the phonetic structure that speech does. There is no set "alphabet" that certain slices of non-speech sound can be split into, and therefore subword features (and the related techniques) cannot be used.

Due to the lack of an environmental sound alphabet, the Hidden Markov Model (HMM) based techniques shown above will be difficult to implement. However, this technique may be revisited in the future if other techniques produce lower than expected results.

In addition, it was decided that the SOM and LVQ techniques compliment each other. Kohonen developed both techniques, with specific applications intended for each technique. For classification, Kohonen suggests the use of the LVQ technique over the SOM technique [9]. Therefore, LVQ will be the technique benchmarked.

Based on this information, the four techniques left to be tested are:

- Dynamic Time Warping
- Long-Term Statistics
- Vector Quantization / Learning Vector Quantization
- Artificial Neural Networks

3. ANALYTICAL ANALYSIS OF SPEECH RECOGNITION TECHNIQUES

This section will detail how each of the techniques listed above were implemented in this system. It will also discuss the details of the experiment (such as number of sounds etc).

3.1 Experiment Setup

As an initial test, eight sounds were used, each with six different samples. Data set size was kept as small as possible due to the time it takes to train larger data sets. The sounds used for this test are detailed below and are some typical sounds that would be classified in a sound surveillance system.

Table 1. Sounds used in Experiment.

Sound Type
Jangling Keys
Footsteps (Close)
Footsteps (Distant)
Wood Snapping
Coins Dropping
Footsteps on Leaves
Footsteps on Glass
Glass Breaking

The techniques will be tested using a jackknife method, identical to the method used by Goldhor [10]. A jackknife testing procedure involves training the network with all of the data except the sound that will be tested. This sound is then tested against the network and the classification is recorded. In cases where the setting of initial weights may affect the classification result (as is the case with LVQ and ANN techniques), classification is repeated 5 times, with different initializations each time. A correct classification is only recorded if more than three of the training runs are correct. This jackknife procedure will be repeated with all six of the samples from each of the eight sounds.

3.2 Benchmarking Method

The feature extraction and system learning techniques shown in the comparison will be tested for their ability to classify non-speech sounds in two ways. First, benchmarking will be performed, using these techniques, on non-speech sounds and data on the parameters, the resulting time taken and the final correct classification rate will be recorded. Then, these results will be compared with statistics and benchmark results reported in the literature for the performance of these techniques on speech. This will demonstrate how these techniques perform against each other on speech and provide a comparison to the results for non-speech.

In addition, since feature extraction and system learning are both required to recognize a sound, each system learning technique should also be tested against each feature extraction technique to determine the best combination of these two techniques. The exception to this is the Long-Term Statistics technique, which generates its own features and therefore requires no feature extraction techniques. Therefore, ten combinations of techniques must be benchmarked:

Table 2. Combination of Feature Extraction/System Learning Techniques.

	Fourier Transform	Mel-Frequency Cepstral Coef.	Homomorphic Cepstral Coef.	Long-Term Statistics
Learning Vector Quantization	✓	✓	✓	
Artificial Neural Networks	✓	✓	✓	
Dynamic Time Warping	✓	✓	✓	
Long-Term Statistics				✓

3.3 Methodology

Each of the techniques used was implemented in MATLAB. Both feature extraction and system learning techniques were implemented and then combined together in the way shown above in order to perform a comprehensive comparison. In this section, the implementation of both the feature extraction and system learning techniques will be discussed.

3.3.1 Feature Extraction Techniques

Three feature extraction techniques will be tested in this comparison. The implementation of each of these techniques will be discussed in this section.

3.3.1.1 Frequency Extraction

Frequency Extraction was performed using the Fast Fourier Transform (FFT) routine in MATLAB, which uses the following equation for FFT:

$$X(k) = \sum_{j=1}^{N} x(j)\omega_N^{(j-1)(k-1)} \qquad k = 1,\ldots,N \qquad (1)$$

with

$$\omega_N = e^{-i2\pi/N}$$

where ω_N is the frequency we wish to check for, j counts all the samples in the signal and N is the length of the signal being tested. Since non-speech sound covers a wider frequency range than speech (anywhere from 0Hz to 20,050Hz, the approximate limit of human hearing), a 44,100 point FFT (N = 44100) was performed and the results (22,050 unique features) were used to train the system learning network.

3.3.1.2 Mel-Frequency Cepstral Coefficients

The MFCC algorithm was taken from the Auditory Toolbox by Malcolm Slaney of Interval Research Corporation [11]. This toolbox is in wide use in the research community. This toolbox applies three steps to produce the MFCC. First, it applies a Hamming Window using the standard Hamming Window equation:

$$w(k+1) = 0.54 - 0.46\cos\left(2\pi\frac{k}{n-1}\right) \qquad k = 1,\ldots,N \qquad (2)$$

where n represents the subset of the signal which is being windowed. A Melody Frequency Filterbank is then applied to each windowed segment. The melody frequency filter bank m is a logarithmic calculation using the following relation:

$$m = \frac{1000\ln\left(1+\frac{f}{700}\right)}{\ln\left(1+\frac{1000}{700}\right)} \approx 1127\ln\left(1+\frac{f}{700}\right) \qquad (3)$$

where f represents the range of frequencies in the signal. Each filter is then multiplied by the spectrum (or portion of the spectrum if it has been split using hamming windows) to produce a series of magnitude values (one for each filter). Finally, a Cepstral Coefficient formula (shown in the next section) is applied to produce MFCC and these features are then modified into a vector that is more appropriate for training a network. Special attention was paid to removing the first scalar within the vector, which represents the total signal power [5] and is therefore too sensitive to the amplitude of the signal [4].

3.3.1.3 Homomorphic Cepstral Coefficients

The MFCC algorithm from the Auditory Toolbox by Malcolm Slaney of Interval Research Corporation [11] was then used as a basis to implement a Homomorphic Cepstral Coefficient (HCC) algorithm. This algorithm was written from scratch but based on information from the source code in the MFCC algorithm.

The HCC algorithm applies the cepstral coefficient formula directly to the signal after it had been split using hamming windows. To calculate cepstral coefficients we use the following relation:

$$y(k) = w(k)\sum_{n=1}^{N} x(n)\cos\frac{\pi(2n-1)(k-1)}{2N} \qquad k = 1,\ldots,N \tag{4}$$

where

$$w(k) = \begin{cases} \sqrt{1/N}, & k = 1 \\ \sqrt{2/N}, & 2 \le k \le N \end{cases}$$

and n is the length of the windowed segment being manipulated. These features were then modified into a vector that was more appropriate for training a network. As with the MFCC, special attention was paid to removing the first scalar within the vector, which represents the total signal power [5] and is therefore too sensitive to the amplitude of the signal [4].

3.3.2 System Learning Techniques

Four system-learning techniques will be tested in this comparison. The implementation of each of these techniques will be discussed in this section.

3.3.2.1 Learning Vector Quantization

Learning vector quantization (LVQ) was implemented using the inbuilt LVQ routines in MATLAB's neural network toolbox. The network was initialized with 20 competitive neurons and a learning rate of 0.05. This combination was found to give an acceptable classification rate.

3.3.2.2 Artificial Neural Networks

Artificial neural network (ANN) was implemented using the fast back propagation algorithm (BPA) in the MATLAB neural network toolbox (*trainbpx*). The network was initialized with 20 hidden neurons and a learning rate of 0.05. In addition, sum-squared error was set to 0.1 and the momentum constant was set to 0.95.

3.3.2.3 Dynamic Time Warping

Dynamic time warping (DTW) was implemented using the algorithm in the Auditory Toolbox developed by Malcolm Slaney [11]. The test signal was warped against each of the reference signals and the error was recorded. The smallest error was taken to represent the closest class of sound.

3.3.2.4 Long-Term Statistics

Long-Term Statistics (LTS) was implemented using the mean and covariance functions available in the standard MATLAB distribution, where N is the length of the signal x. Mean and covariance were calculated for each of the reference signals and stored in a matrix. The mean and covariance of the test signal was then compared to this matrix. The closest match was selected as the correct class. If the closest mean and covariance occurred in difference classes, the test was concluded to be inconclusive.

4. RESULTS & DISCUSSION

This section will cover the results of this research. Results are shown for the comparative study of existing speech recognition techniques when these techniques are applied to non-speech. In addition, a discussion is given on these results.

4.1 Results

4.1.1 Non-Speech Sound Recognition

Results for non-speech sound recognition are presented below.

Table 3. Learning Vector Quantization (LVQ) for Non-Speech Sound Recognition.

	150 Epochs		300 Epochs		450 Epochs	
Method	Time	% Correct	Time	% Correct	Time	% Correct
FT	35 sec	41%	42 sec	58%	64 sec	63%
MFCC	2 sec	33%	3 sec	46%	6 sec	58%
HCC	1 sec	17%	2 sec	29%	4 sec	33%

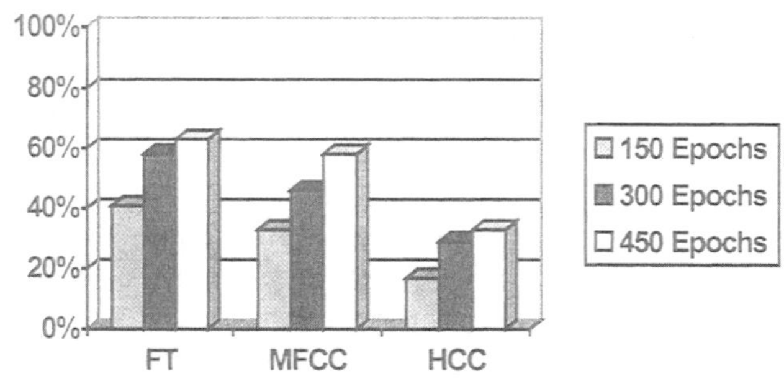

Figure 1. Comparison of LVQ for Non-Speech Sound Recognition.

Table 4. Artificial Neural Network (ANN) for Non-Speech Sound Recognition.

	150 Epochs		300 Epochs		450 Epochs	
Method	Time	% Correct	Time	% Correct	Time	% Correct
FT	145 sec	8%	324 sec	8%	483 sec	13%
MFCC	5 sec	8%	16 sec	8%	30 sec	8%
HCC	3 sec	0%	7 sec	8%	13 sec	8%

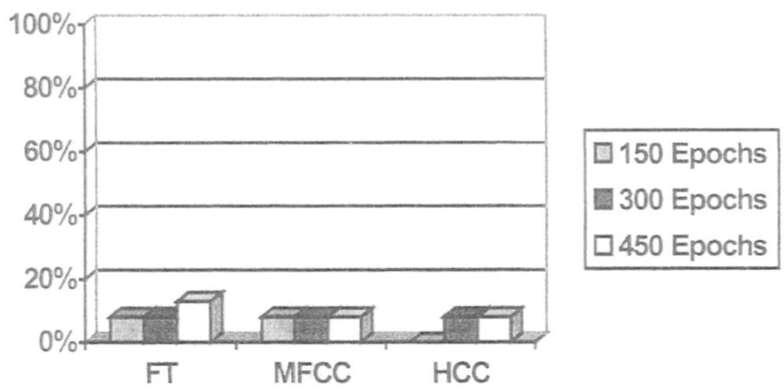

Figure 2. Comparison of ANN for Non-Speech Sound Recognition.

Table 5. Dynamic Time Warping (DTW) for Non-Speech Sound Recognition.

Method	Time	% Correct
FT	< 1 sec	66%
MFCC	< 1 sec	70%
HCC	< 1 sec	29%

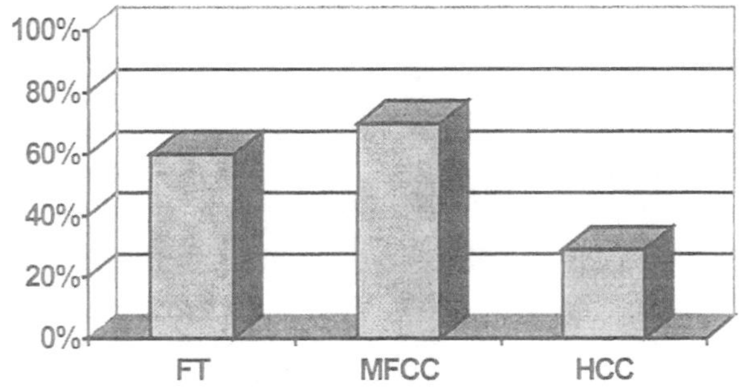

Figure 3. Comparison of DTW for Non-Speech Sound Recognition.

Table 6. Long-Term Statistics (LTS) for Non-Speech Sound Recognition.

Method	Time	% Correct
FT	< 1 sec	29%
Power FT	< 1 sec	29%

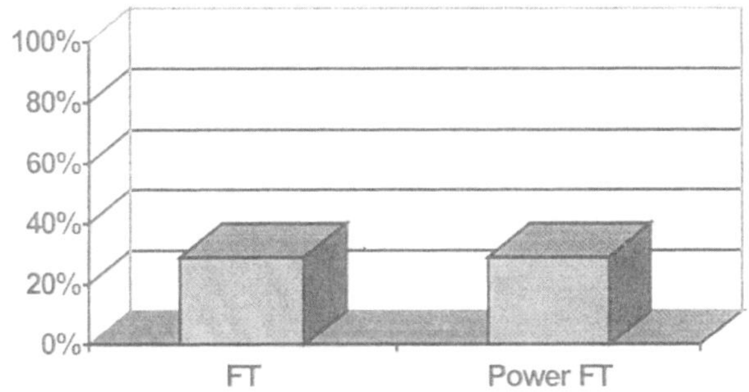

Figure 4. Comparison of LTS for Non-Speech Sound Recognition.

4.1.2 Speech Recognition

For comparison, results were found for LVQ and ANN in speech recognition systems. These results are presented here. Due to the current popularity of HMM methods in speech recognition at the present time, results for DTW are difficult to find, therefore no DTW results are presented.

For ANN's, a selection of results from Castro and Perez [12] are shown below. Their results were taken on an isolated word recognition set with typically high classification error, the Spanish EE-set. The Multi-Layer Perceptron (MLP) tested used the back propagation algorithm, contained 20 hidden neurons and was trained over 2000 epochs with various amounts of inputs. The figures given are the MLP's estimated error rate with a 95% confidence interval.

For LVQ, results from Van de Wouver e.a. [13] are shown below for both female and male voices. These results present statistics for both a standard LVQ implementation for speech recognition and an implementation of LVQ that then has fuzzy logic performed on it (FILVQ). As can be seen from the results, the use of LVQ for speech recognition produces rather low recognition results.

Table 7. ANN for Speech Recognition [12].

Number of Inputs	% Correct
550 inputs	80.3%
220 inputs	83.7%

Table 8. LVQ for Speech Recognition [13].

Method	% Correct (Female)	% Correct (Male)
Standard LVQ	36%	29%
FI-LVQ	60%	64%

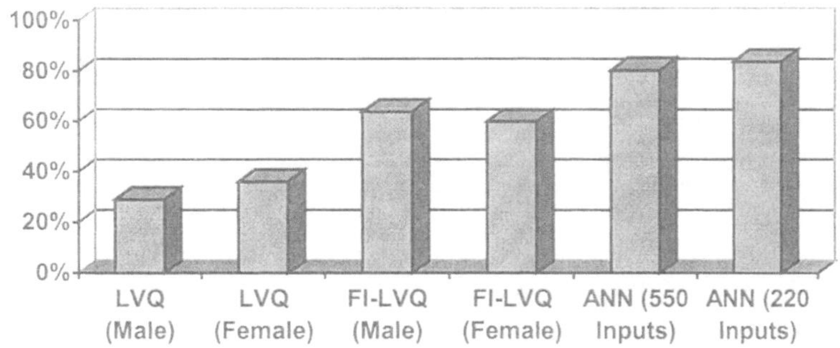

Figure 5. Comparison of Speech Recognition Results.

4.2 Discussion

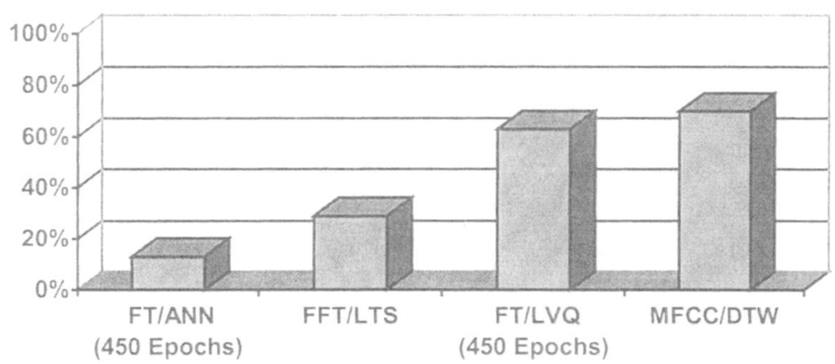

Figure 6. Comparison of Best Results for Non-Speech Sound Recognition.

The results obtained are somewhat surprising. Even though the results from speech recognition suggest that the ANN will outperform LVQ, the opposite occurs for non-speech recognition. We propose that this is due to the closeness of the various environmental sounds presented to the two networks.

It is accepted that one of the main advantages of LVQ over ANN is its ability to correctly classify results even where classes are similar. In this case, sounds such as footsteps (close) and footsteps (distant) appear the same but contain slightly lower or higher amplitudes. LVQ is able to classify these sounds properly where the ANN cannot distinguish them. Furthermore, the detailed results of each test show that the ANN was classifying footsteps (close) as footsteps (distant) and vice versa. To prove this hypothesis, further tests were performed on the ANN using several higher epoch values (to allow more training time). The results are presented below.

Table 9. Further ANN for Non-Speech Sound Recognition.

	200 Epochs		800 Epochs		1200 Epochs	
Method	Time	% Correct	Time	% Correct	Time	% Correct
FT	184 sec	0%	742 sec	1%	1087 sec	13%
MFCC	5 sec	0%	38 sec	0%	38 sec	4%

These results are graphed below. From these results, it can be seen that the ANN results remain the same regardless of the epoch value. This suggests that the ANN has problems training the sample sounds, most likely due to these sounds being non-linearly separable.

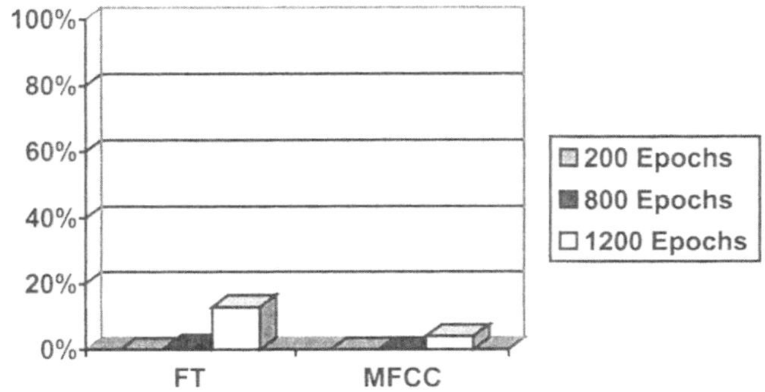

Figure 7. Comparison of Further ANN for Non-Speech Sound Recognition.

The performance of the Mel Frequency Cepstral Coefficient (MFCC) feature extraction algorithm over the Fourier Transform (FT) Based Frequency Extraction algorithm is also interesting. Surprisingly, in all cases except the when DTW is used as a classifier, the FT algorithm outperforms the MFCC algorithm by approximately 5% on 450 epochs and approximately 10% on 300 epochs. However, in order for the FT algorithm to achieve these results, it has to spend almost 10 times as long training as the MFCC algorithm does.

In addition, for the LVQ and ANN tests, it seems that the MFCC algorithm can achieve a maximum classification rate of approximately 58%, with only slight variations dependant on epochs. In contrast, the FT algorithm can achieve a slightly higher rate, reaching its maximum at around 65% (further testing has shown that more epochs do not improve this classification rate).

The DTW algorithm also produces surprising results. This algorithm shows only a small difference (equivalent to one classification) in performance between the MFCC algorithm and the FT algorithm. This is in contrast to the large difference between these algorithms in the LVQ and ANN tests. In addition, DTW performs classification much quicker than the LVQ and ANN techniques. This is most likely due to the fact that DTW does not require any training and instead relies on a series of reference models. However, the downside to this approach is the extra storage space required for these templates.

Finally, in contrast to results presented by Lilly [4], these results show a substantial difference in classification between the HCC technique and the MFCC technique. Due to the fact that other researchers report similar classification rates using these two techniques, implementation of these techniques could conceivably be improved. However, since MFCC seems to be the more popular technique and produces the better results of the two techniques, at this stage it will continue to be used in its current form.

In addition, time-frequency techniques also present another avenue of investigation, in the same way as they are currently being investigated for speech and speaker recognition [14].

5. CONCLUSION

This chapter has presented the results of benchmarking several common stationary signal feature extraction techniques for speech on environmental sounds. Based on these results, the most desirable technique would seem to be a combination of a MFCC based feature extraction technique with a DTW based system learning technique. This combination produces the best classification rate, followed by the use of a FT feature extraction technique with a DTW system learning technique.

Further research should make systematic comparisons of other classification techniques from speech recognition that could be used for non-speech sound classification. Specifically, time-frequency techniques such as short-time Fourier transform (STFT) and wavelets should be investigated for non-speech, in the same way as they are currently being investigated for speech and speaker recognition.

REFERENCES

[1] R. Brooks, C. Breazeal, M. Marjanovic, B. Scassellati, and M. Williamson, "The Cog Project: Building a Humanoid Robot," C. Nehaniv, ed., *Computation for Metaphors, Analogy and Agents*, Springer Lecture Notes in Artificial Intelligence, Springer-Verlag, Vol. 1562, 1998

[2] M. Cowling, R. Sitte, "Sound Identification and Direction Detection in MATLAB for Surveillance Applications," *Proc. MATLAB Users Conference*, Melbourne, Australia, November 2000.

[3] M. Cowling, R. Sitte, "Sound Identification and Direction Detection for Surveillance Applications," *Proc. of ICICS 2001*, Singapore, October, 2001.

[4] B. Lilly, *Robust Speech Recognition in Adverse Environments*, PhD Thesis, Faculty of Engineering, Griffith University, Nathan Campus, May 2000.

[5] B. Gold, N. Morgan, *Speech and Audio Signal Processing*, John Wiley & Sons, Inc, 2000, New York, NY.

[6] C. H. Lee, F. K. Soong, K. Paliwal, "An Overview of Automatic Speech Recognition," in *Automatic Speech and Speaker Recognition: Advanced Topics*, Kluwer Academic Publishers, 1996, Norwell, MA.

[7] C. H. Lee, F. K. Soong, K. Paliwal, "An Overview of Speaker Recognition Technology," in *Automatic Speech and Speaker Recognition: Advanced Topics*, Kluwer Academic Publishers, 1996, Norwell, MA.

[8] R. Rodman, *Computer Speech Technology,* Artech House, Inc. 1999, Norwood, MA 02062.

[9] T. Kohonen, *Self-Organizing Maps,* 1997, Springer-Verlag Berlin, Germany.

[10] R. S. Goldhor, "Recognition of Environmental Sounds," *Proc. ICASSP,* Vol. 1, pp 149 – 152, New York, NY, USA, April 1993

[11] M. Slaney, "Auditory Toolbox"™, Interval Research Coporation, version 2, 1998.

[12] M. J. Castro, J. C. Perez, "Comparison of Geometric, Connectionist and Structural Techniques on a Difficult Isolated Word Recognition Task," *Proc. European Conference on Speech Comm. and Tech., ESCA*, Vol. 3, pp 1599-1602, Berlin, Germany, 1993.

[13] G. Van de Wouver, P. Scheunders, D. Van Dyck, "Wavelet-FILVQ Classifier for Speech Analysis," *Proc. Int. Conference Pattern Recognition*, pp. 214-218, Vienna, 1996.

[14] M. Orr, D. Pham, B. Lithgow, R. Mahony, "Speech perception based algorithm for the separation of overlapping speech signal," *Proc. The Seventh Australian and New Zealand Intelligent Information Systems Conference*, pp. 341 - 344 Perth, Western Australia, 2001.

Chapter 4

A NOVEL DUAL ADAPTIVE APPROACH TO SPEECH PROCESSING

Michael C. Orr, Brian J, Lithgow, Robert Mahony, Duc Son Pham
Monash University Centre for Biomedical Engineering, Monash University, Melbourne, Australia

Abstract: This paper presents a wavelet transform based analysis technique for speech separation of overlapping speakers. A continuous Morlet wavelet (equation 1) [5] was used for analysis since it reputedly has the best time-frequency localization (i.e. closest to Gaussian in both the time and frequency domains) of all wavelets [18]. The extracted wavelet coefficients are analyzed using a covariance matrix to provide information about speaker's characteristics. Second a statistical analysis for extracting some phonetic information, such as articulation placement and identification of voiced/unvoiced sections of speech.

Key words: Channel identification, digital signal processing, DSP, Morlet, Continuous wavelet transform, CWT, Eigen space, wavelet space

1. INTRODUCTION

Many speech-processing techniques try to mimic the auditory filter bank (organ of Corti). Usually solely based on the signal's spectrum, e.g. cepstral analysis [17], and assume the signal is stationary. However, speech and many environmental sounds are time varying signals. Human hearing utilizes both spectral and temporal information to filter sounds in a noisy environment [19]. Time-frequency techniques can utilize variable time windows for improved time resolution. As such, they are better able to localize time varying signals than Fourier analysis.

Building on ongoing work in the design of hearing aid systems, the proposed technique (figure 1) has a wavelet and inverse wavelet transform to enable analysis in the time frequency domain. Adaptation algorithms built around the transforms have the dual purpose of compensating for distortion and noise in the input channel, as well as filtering for specific speech features. A second adaptive procedure operates on a much slower time scale, periodically updating the particular wavelet templates used in the transform to account for changes in the environment. In particular, the optimization / identification algorithm developed can be tuned to optimize the transform for different speakers.

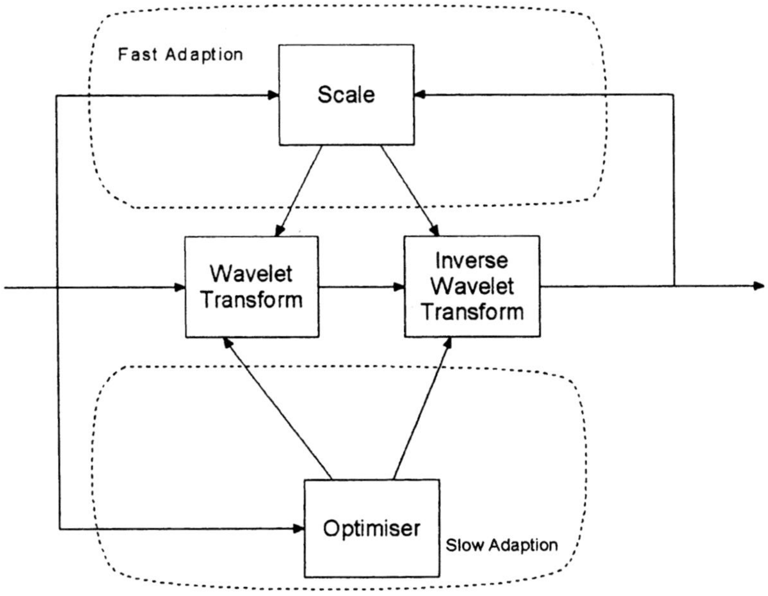

Figure 1. Novel dual adaptive scheme for speech processing (reproduced with permission from [13])

The authors hypothesize, templates of time-frequency characteristics of sounds, whether speech or environmental, exist both in the short and long term (figure 2). Anatomical and physiological differences between people's vocal tracts mean every person's voice is different. However, algorithmic characterization/identification of vocal tract parameters requires analysis of continuous speech over time. Instantaneous analysis of the anatomical and physiological parameters of speech is not possible, since the full range of vocal tract configurations do not appear in every sound. Accordingly the speaker characteristic block, in figure 2, extracts and adapts, over a period of tens of seconds, time-frequency templates of the vocal-tract. Nevertheless, due to changes in the environment, or other reasons, a person's voice will sound different over a short period (a few seconds). The linguistic characteristic block, in figure 2, extracts and adapts, over a short period, time-frequency templates of recent sounds, both speech and environmental.

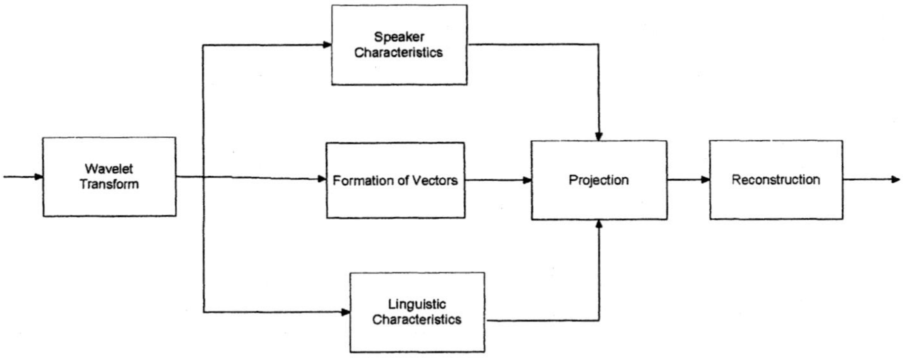

Figure 2. Feature extraction of both long and short-term time-frequency templates, reproduced with permission from [13].

The work presented in this chapter is an overview of one method of wavelet analysis for speech. Those readers interested in an explanation of the classical methods for speech processing should read Michael Cowling's et al chapter titled "Recognition of Environmental Sounds using Speech Recognition Techniques".

2. WAVELET TRANSFORM FILTER BANK

The two main considerations in designing a wavelet filter bank are the bandwidth coverage and separation of filters. A major difference between implementation of a Morlet filter bank and a Fourier based filter bank, is that Fourier based filters have linear frequency scale whilst the Morlet filter bank has a logarithmic frequency scale. Hence, implementation of a Morlet filter bank requires a quality factor to be determined. A reasonable reduces the spreading of formant frequencies across filters. A 3dB cut-off point was chosen for minimum correlation between filters whilst maintaining a reasonably flat response across the spectrum. Appendix A contains the derivation for determining the number of filters per octave with a 3dB cut-off overlap and arbitrary quality factor. Using the derived formula it was determined four filters per octave with a quality factor of 5.875 were required for capturing of speech features. For the purposes of analysis, the frequency content in a speech signal was band limited to between 100Hz and 8kHz. Table 1 shows the frequency range of the filter bank.

$$\psi = \frac{e^{\frac{-t^2}{2}} e^{j\omega t}}{\sqrt{2\pi}} \tag{1}$$

Table 1. Frequency range of filter bank

Octave	Low 3dB frequency	High 3dB frequency
1	62.5 Hz	125 Hz
2	125 Hz	250 Hz
3	250 Hz	500 Hz
4	500 Hz	1 kHz
5	1 kHz	2 kHz
6	2 kHz	4 kHz
7	4 kHz	8 kHz

Tests showed the wavelet filter bank had an acceptably flat frequency response and ability to reconstruct clean speech [14]. Perceptually there was little difference between the original speech and the reconstructed speech. A slight change in the magnitude of the frequency response within octaves was detected and determined to be due to sampling of the Morlet wavelet.

3. SPEAKER SPACE EXPERIMENTS

Initial work has concentrated on characterizing different speaker vocal tracts, via a time-frequency signal space, over a period of tens of seconds. After analyzing speech (of a single speaker) using a continuous wavelet transform (CWT), a covariance (or correlation matrix) of the time-frequency information, was formed. Eigen vector decomposition of the correlation matrix was used to identify a subspace of wavelet coefficient space for a given speaker vocal characteristics. This space referred to as a "speaker space" represents a finite basis of linear combinations of filter bank coefficients. If the wavelet bank was designed with sufficient fidelity to uniquely represent a given speakers voice then the difference between speakers can be measured in terms of the angle between their representative subspaces [16].

Figure 3 is a plot of eigen value vs. degree of separation for three distance measurements. The authors have shown separation of 45 degrees is equivalent to 3dB attenuation [13]. Eigen vectors separated by more than 60 degrees indicate the system is able find distinguishing features between speakers [13]. Vectors with an angular difference of less than 10 degrees suggest the system is representing noise [13]. Vectors in the third region i.e. between 10 and 60 degrees apart represent similarities between speakers [13].

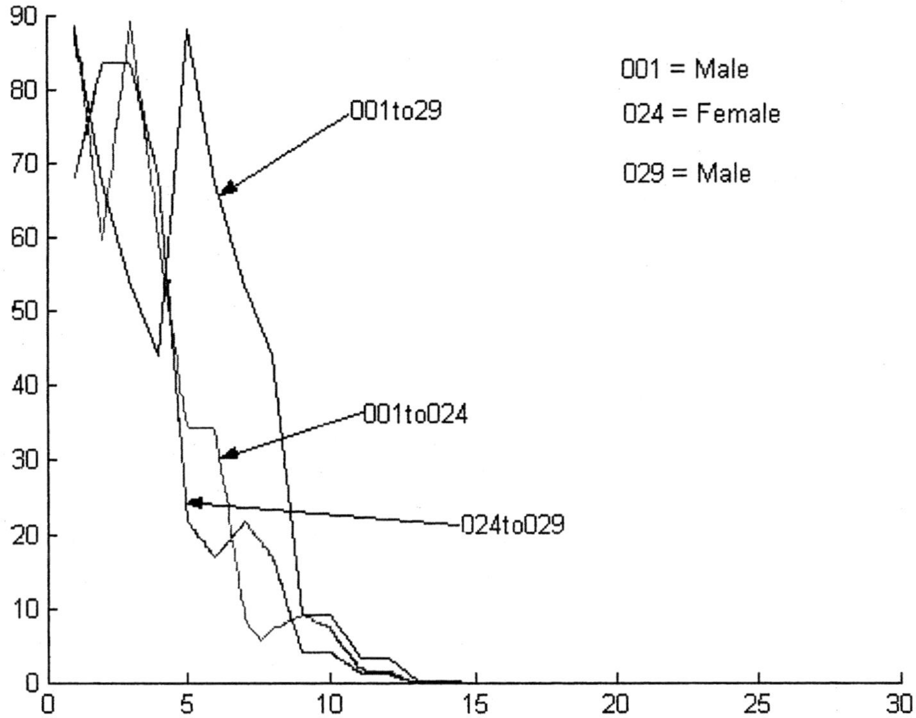

Figure 3. Measurement of angular difference between speaker spaces

It was obvious from figure 3 that the eigen values with an index greater than 13 are most likely characterizing random noise due to the recording process and not speech. Therefore reducing the speaker space dimensionality by not including these dimensions will allow reconstruction with a reduced level of noise corruption. Although, it is well established that principal component analysis (PCA) does not efficiently separate correlated signals [7, 2], the authors considered it necessary to establish a baseline speech separation result before continuing onto more complex algorithms for dimension reduction. Dimension reduction experiments using PCA also allowed the characterization system to be tested. Principal component analysis essentially optimizes the variance of a signal [3, 6]. After constructing several speaker spaces using the ANDSOL library of speech files [11] each space had its dimensions reduced by selecting on the largest eigen values i.e. principal component analysis. Next, corrupted speech containing a "target" speaker and another speaker (not necessarily one of the other speakers characterized) was analyzed using a continuous wavelet transform. After projecting the vectors of wavelet coefficients for each sample, of corrupted speech, onto the "target" speaker space, an inverse wavelet transform reconstructed the speech. Statistically the results achieved indicated the need for two improvements, an algorithm(s) to detect segments speech containing only one speaker and the ability to detect the onset and duration of unvoiced speech.

4. STATISTICAL MEASURES OF SPEECH

In many situations with random signals, and random environmental parameters, a statistical approach has proven to be a powerful tool. Furthermore, higher-order statistics (such as kurtosis, skew) and associated methods have been recognized as essential and efficient statistics in problems of blind equalization. While such higher order moments require a large number of samples for accurate estimation, many engineering applications implement algorithms in which their rough estimates for non-stationary signals, e.g. the human voice, provide enough accuracy for successful processing in the speech separation problem.

Kurtosis is a statistical measure of the shape of a distribution [1]:

$$Kurtosis = \frac{\sum (X-\mu)^4}{N\sigma^4} \tag{2}$$

where N is the number of data points in the distribution, σ is the standard deviation, μ is the mean, and X is the data.

A Gaussian signal has a kurtosis of three [1]. From an engineering perspective, kurtosis measures the distribution of energy across a signal window. If the energy is concentrated in tight packets, one or many, then the kurtosis is less than three. However, the flatter the energy within a window the greater the kurtosis above three. Le Blanc [9] proposed speech has a flatter than Gaussian energy distribution across time resulting kurtosis values of greater than three. Contradicting this work Nemer [12], using a sinusoidal model for speech [10], proposed that the kurtosis of (voiced) speech is less than three. Results presented in the next section (figure 4) show that the kurtosis of speech is below three during voiced speech.

Skewness is a statistical measure of the distribution of data across a window [8]:

$$Skew = \frac{\sum (X-\mu)^3}{N\sigma^3} \tag{3}$$

where N is the number of data points in the distribution, σ is the standard deviation, μ is the mean, and X is the data. A normal distribution has zero skew. Distributions whose mean is less that the median have a negative skew and distributions with a mean higher than the median have a positive skew [8].

5. USING STATISTICAL MEASURES OF SPEECH TO DETECT LINGUISTIC FEATURES

In a preliminary study of fifteen phonetically significant sentences, each spoken by four speakers, two males and two females were analyzed using kurtosis and skew measures. All the sentences used in the study are from the ANDSOL database [11]. The files for each speaker were analyzed using MATLAB®, a sliding window of five milliseconds applied in the time domain with 99 percent overlap. A five-millisecond window is appropriate for identifying short-term features, for example stop consonants, in speech. Simultaneously the files were also annotated to isolate four sounds of interest, the fricative pair /v/ and /f/, as well as stop consonant pair /g/, and /k/. In total four examples of each sound were isolated in the fifteen sentences.

Figure 4 and figure 5 show kurtosis and skew analysis, respectively, of a male speaker saying /v/ in the word "never". Subjective analysis of the kurtosis and skew curves show that periods of voiced speech, unvoiced speech, and silence, have identifiable characteristics. For example voiced sections of /v/ are identifiable in sections with zero average skew.

Preliminary analysis shows consistent identifiable features for both kurtosis and skew across utterances and across speakers. Through a process of comparing the annotated sounds with corresponding kurtosis and skew data, it was possible to isolate several identifiable and consistent features. Voiced sections of speech have a kurtosis of between one and two and a skew close to zero. In general, unvoiced sections of speech can be characterized by large magnitude of both kurtosis and skew. Silence periods are the hardest to distinguish but a rule of thumb is a kurtosis similar to voiced speech and a skew similar to unvoiced speech.

The curves for kurtosis and skew like speech are subjectively different between utterances or speakers for the same sound. Coefficient magnitudes for kurtosis and skew are dependent on the frequencies present in the speech and the size of the analysis window. Analysis of the waveforms failed to provide an objective set of rules for distinguishing features. Relative magnitudes across time seemed to give a better indication than actual magnitudes at a single instant.

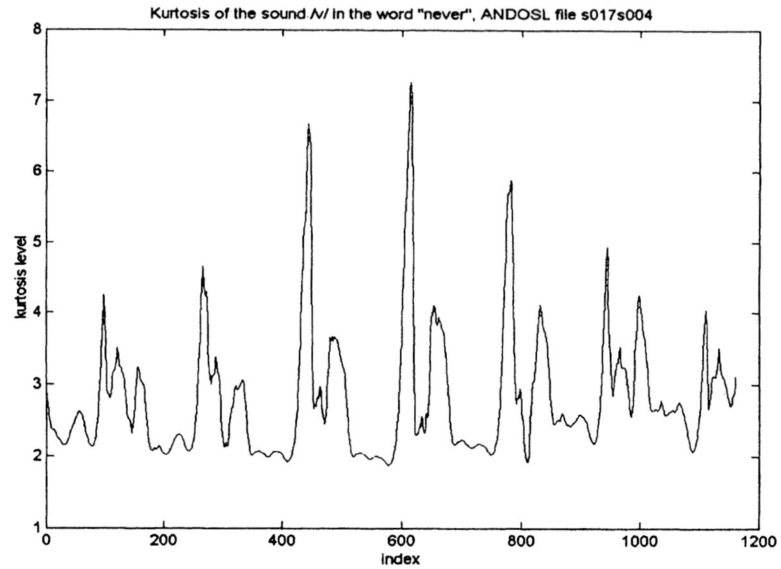

Figure 4. Kurtosis of the sound /v/ in the word "never"

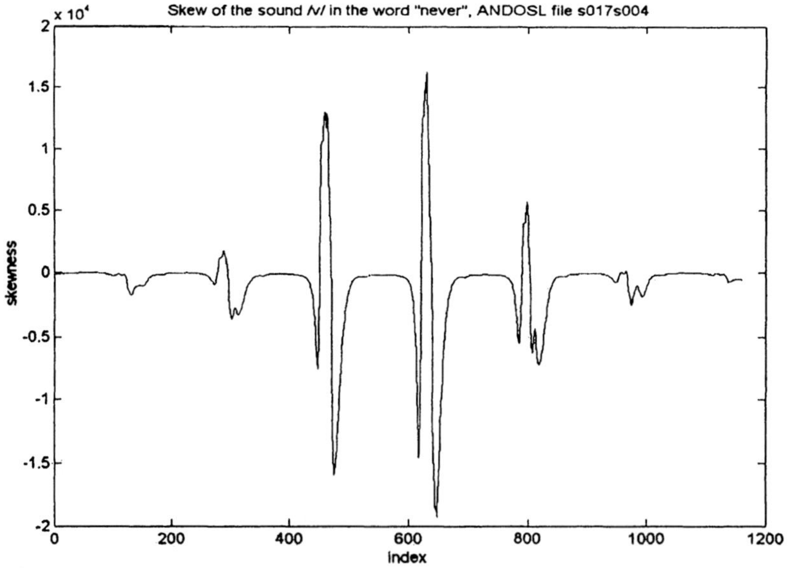

Figure 5. Skew of the sound /v/ in the word "never"

6. CONCLUSION

The novel dual adaptive approach to speech processing outlined (Figure 1) in this chapter is equally applicable to either communication or medical systems. It is hypothesized, by the authors, that an on-line dual adaptation and optimization based channel identification scheme provides potential for improvements to speech processing algorithms. Preliminary investigations of the approach have concentrated on improving feature extraction for speaker separation, noise reduction, and qualitative improvement in hearing aid performance in difficult situations such as encountered with the cocktail-party effect [15]. Envisaged applications to other medical problems include image processing and extraction of individual biological signals from a montage. Other, possible applications include speaker identification, security, communications, word recognition, image processing, and signal separation.

The best basis selection criteria are being developed to incorporate the use of statistical measurements in a similar manner to blind equalization. The wavelet filter bank is under review with the objective of better emphasizing the time-frequency features of speech, particularly the difference between speakers.

APPENDIX A: FILTER BANK PROOF

Relationship between the Q of the filter bank and the center frequencies of the filters

$$Q = \frac{f_{c_1}}{BW_1} = \frac{f_{c_2}}{BW_2} \tag{4}$$

$$BW_2 = \frac{BW_1 f_{c_2}}{f_{c_1}} \tag{5}$$

Assuming a cut-off point of 3dB

$$f_{c_2} = f_{c_1} + \frac{BW_1}{2} + \frac{BW_2}{2} \tag{6}$$

Substituting (4) and (5) into (6) yields

$$f_{c_2} = f_{c_1} + \frac{BW_1}{2} + \frac{BW_1 f_{c_2}}{2 f_{c_1}} \tag{7}$$

$$f_{c_2} = 2Q\left[\frac{f_{c_1} + \frac{f_{c_1}}{2Q}}{2Q-1}\right] \tag{8}$$

$$f_{c_2} = f_{c_1}\left[\frac{2Q+1}{2Q-1}\right] \tag{9}$$

On a logarithmic frequency scale f_{c_2} and f_{c_1} are separated by

$$\log f_{c_2} - \log f_{c_1} = \alpha(\log 2) \tag{10}$$

where α is the separating fraction of an octave.
Hence

$$\text{Octave Separation of filters} = \frac{\log\left[\frac{2Q+1}{2Q-1}\right]}{\log 2} \tag{11}$$

We want one filter coefficient to be wide enough to capture an F0 at 180 hertz. The bandwidth of F0 is approximately 30 hertz. Therefore, the required Q is equal to 6.

Substituting Q = 6 into (11) yields

$$\text{Octave Separation of filters} = 0.241$$

For simplicity, we set the Octave Separation to 0.25, giving four filters per octave. Therefore, the Q of each filter is approximately 5.7852.

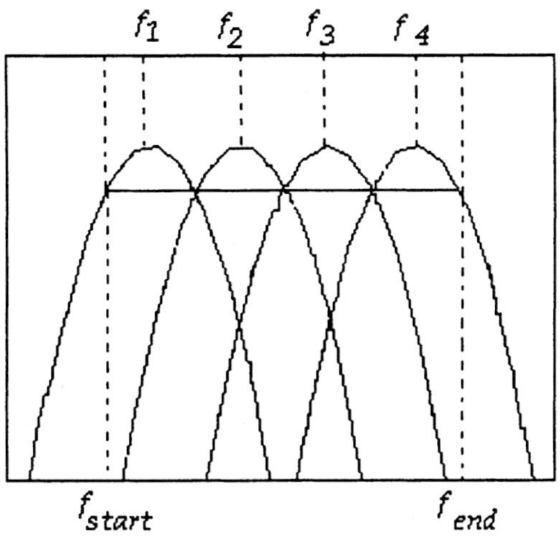

Figure 6. Four filters per octave

To extend the design of the filter bank to a whole octave we need to apply (7), (8) and (9) to the two filters. As a result, we get:

$$f_{c_3} = f_{c_2}\left[\frac{2Q+1}{2Q-1}\right] \tag{12}$$

$$f_{c_4} = f_{c_1}\left[\frac{2Q+1}{2Q-1}\right]^3 \tag{13}$$

Hence, to maintain the 3 dB separation between octaves

$$f_{c_1} = 1.095 f_{start}$$

$$f_{c_2} = 1.3017 f_{start}$$

$$f_{c_3} = 1.548 f_{start}$$

$$f_{c_4} = 1.8409 f_{start}$$

REFERENCES

[1] Bulmer, Michael G., *Principles of statistics*, 1st ed: Oliver & Boyd, 1965.

[2] Chang, Wei-Chen, "On using principal components before separating a mixture of two multivariate normal distributions," *Applied Statistics*, vol. 32 (3) pp. 267 - 275, 1983.

[3] Dunteman, George H., Principle components analysis, *Quantitative applications in the social sciences.*: SAGE Publications, 1989.

[4] Fant, G., *Acoustic theory of speech production, 1st ed*: Mouton & Co., 1960.

[5] Goupillaud, P., Grossmann, A., and Morlet, J., "Cycle-octave and related transforms in seismic signal analysis," *Geoexploration*, vol. 23 pp. 85 - 102, 1984.

[6] Jolliffe, I. T., *Principal component analysis*. New York: Springer-Verlag, 1986.

[7] Kambhatla, Nandakishore and Leen, Todd K., "Dimension reduction by local principal component analysis," *Neural Computation*, vol. 9 pp. 1493 - 1516, 1997.

[8] Lane, Daivd M., "*Hyperstat*" at http://davidmlane.com/hyperstat/index.html downloaded 20th March 2002 last updated 17th March 2000

[9] Le Blanc, James P. and De Leon, Phillip L., "Speech separation by kurtosis maximization," in Proc. of *IEEE International Conference on Acoustics, Speech and Signal Processing*, 1998.

[10] Mcaulay, Robert J. and Quatieri, Thomas F., "Speech analysis/synthesis based on sinusoidal representation," *IEEE Transactions on Acoustics, Speech, and Signal Processing*, vol. 34 (4) pp. 744 - 754, 1986.

[11] Millar, J.B., Vonwiller, J.P., Harrington, Jonathan .M., and Dermody, P.J., "The australian national database of spoken language," in Proc. of *IEEE International Conference on Acoustics, Speech, and Signal Processing, ICASSP-94*, Adelaide, SA, Australia, 1994.

[12] Nemer, Elias, Goubran, Rafik, and Mahmoud, Samy, "Speech enhancement using fourth-order cumulants and optimum filters in the subband domain," *Speech Communication*, vol. 36 (3) pp. 219 - 246, 2002.

[13] Orr, Michael C., "Progress report: Speech separation," Monash University, *Quarterly report submitted to ECSE*, September 2001.

[14] Pham, Duc Son, Orr, Michael C., Lithgow, Brian J., and Mahony, Robert E., "A practical approach to real-time application of speaker recognition using wavelets and linear algebra," in Proc. of *6th International Symposium on DSP for Communication Systems*, Manly, NSW, Australia, 2002.

[15] Steinberg, J.C. and Gardener, M.B., "The dependence of hearing impairment on sound intensity," *Journal of the Acoustical Society of America*, vol. 9 pp. 11 - 23, 1937.

[16] Stewart, G.W., "Error and perturbations bounds for subspaces associated with certain eigenvalue problems," *SIAM Review*, vol. 15 (4) pp. 727 - 764, 1973.

[17] Tokuda, Keiichi, Kobayashi, Takao, and Imai, Satohsi, "Adaptive cepstral analysis of speech," *IEEE Transactions on Speech and Audio Processing*, vol. 3 (6) pp. 481 - 489, 1995.

[18] M. Unser, and A.Aldroubi, "A review of wavelets in biomedical applications," *Proceedings of the IEEE*, vol. 84 (4) pp. 626 - 638, 1996.

[19] Zwicker, E and Fastl, H, *Psychoacoustics: Facts and models*, 2nd ed., Springer-Verlag, 1999.

Chapter 5

ON THE DESIGN OF WIDEBAND CDMA USER EQUIPMENT (UE) MODEM

KyungHi Chang, MoonChil Song, HyeongSook Park, YoungSeog Song, Kyung-Yeol Sohn, Young-Hoon Kim, Choong-Il Yeh, ChangWahn Yu, and DaeHo Kim

ETRI

Abstract: This paper describes the details of the implementation issues on wideband CDMA UE Modem, concentrating on the receiving side. Employed top-down design methodology and corresponding work efforts on each task are stated to demonstrate the effectiveness of the methodology.

Key words: WCDMA, 3GPP, UE Modem

1. INTRODUCTION

Design methodology and the technical details of 3GPP Compliant UE Modem, which has been implemented based on [1]-[8], are treated in this paper. Basically, this Modem has been tested including following advanced features :

- Tx Diversity
  - Open-Loop Diversity : Space Time Transmit Diversity (STTD) & Time Switched Transmit Diversity (TSTD)
  - Closed-Loop Diversity : Mode 1 & 2
- Power Control (PC)
  - Inner-Loop PC
    - Open-Loop PC
    - Closed-Loop PC
      - Down-Link Power Control (DL PC)

        - Up-Link Power Control (UL PC)
    - Outer-Loop PC
    - Site Selection Diversity Transmit Power Control (SSDT)

- Compressed Mode (CM)
- includes AICH, PICH, CPCH-related Indicate Channels (ICHs), PDSCH and CPCH.
- supports upto 3 Multi-Codes (MCs) in DL DPCH and 6 MCs in UL DPCH.

2. DESIGN METHODOLOGY

From the specifications to the test of the implemented system, tasks are categorized as follows :

- Task 1 : Specifications & Requirement Analysis
- Task 2 : Floating-Point Modeling
- Task 3 : Fixed-Point Design
- Task 4 : VHDL & SW Implementation
- Task 5 : Prototyping & Test

Task 1, Specifications & Requirement Analysis, denotes the overall system specifications in terms of functionality which might be implemented and the requirement analysis of the target hardware, e.g., FPGA & ASIC, u-Controller and analog front-end (AFE) etc. Verification and test plan of successive tasks can also be developed during this task.

Task 2, Floating-Point Modeling, means the reference functional model development. This task should be achieved by considering the performance and complexity simultaneously to select the potential algorithm. High-level design tool could be adopted to ease the overall design and verification process.

Task 3, Fixed-Point Design, is the development of the architectural model. Output of this task, i.e., fixed-point design, is compared with the floating-point model in terms of performance with determined bit-width of design. HW & SW partitioning should be done within this task. Exactness of fixed-point design and VHDL implementation can be managed depending on the work efforts of design & test stage. Same design environment as in Task 2 might reduce the design efforts in Task 3.

Task 4, VHDL & SW Implementation, is the task of implementation regarding HW & SW. Design environment in terms of top-down solution might help the overall approach. Various Co-Simulation strategy for the fixed-point model & VHDL model and/or the HW & SW could be set up. In case of SW implementation, requirements of SW task and interfaces with HW should be carefully treated.

Task 5, Prototyping & Test, includes prototype system design, such as FPGA/ASIC-based system, and functional & performance test of the prototype system. Development of the appropriate test environments is also the part of this task. This test phase can be divided into stand-alone and system-level integrated test.

During the overall design process, any feedback to the previous tasks should always be allowed.

Work efforts among the tasks on the design of the wideband CDMA UE modem have roughly been as follows. Task1 : Task2 : Task3 : Task4 : Task5 = 2 : 1 : 3 : 3 : 6.

3. UE DEMODULATOR

We focus our discussion on the downlink receiving side first, i.e., UE demodulator. Top-level architecture of UE demodulator is shown in Figure 1.

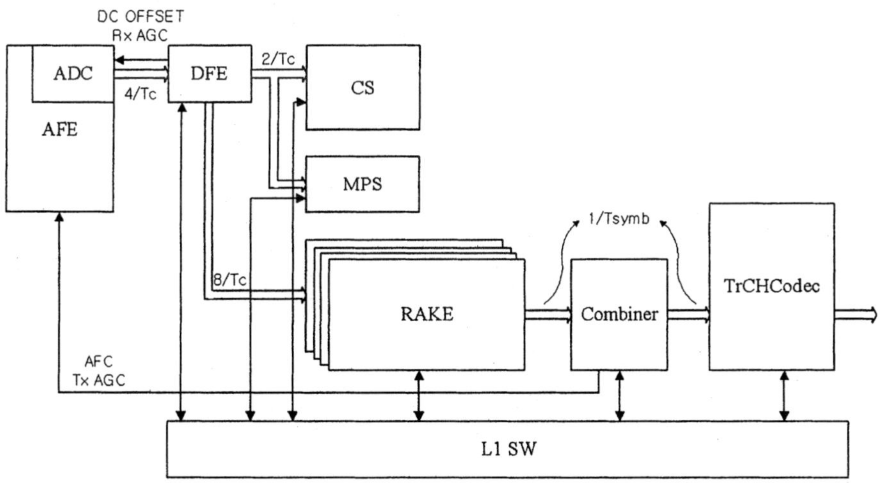

Figure 1. Top-level architecture of UE demodulator.

3.1 Digital Front-End (DFE)

DFE consists of matched filter (MF), down-sampling to appropriate rate by decimation, up-sampling to 8/Tc by interpolation, DC offset compensation and DAGC. The ADC output signal, which is provided at 4/Tc, is filtered by the 40-tap digital matched filter. This front-end decimation filter let the matched filter consume less power. The matched filter output rate is reduced by a factor of 4 to 1/Tc by a sampling device and provided as an input of DC offset compensator. A DC offset on the matched filter output is removed by the DC offset control loop. The DC offset at the matched filter output (resulting from analog DC voltages, AD-conversion and truncation in the matched filter) is computed by the offset compensation block. To control the receiving power of UE, the power at the matched filter output, after decimating by a factor of 2 to 2/Tc, is monitored by the DAGC. The DAGC generates a sigma-delta modulated control signal for an external gain controlled amplifier to keep the matched filter (and front-end) output power at a constant average level. The control signal for the analog offset compensation circuits is provided via sigma-delta modulation to close the mixed analog/digital offset compensation loop. The DAGC error level and the power level are provided as outputs, to be read by the u-Processor to derive input power measurements. This 2-times decimated matched filter output also becomes the input of cell searcher (CS) and multipath searcher (MPS).

An interpolation filter increases the sampling rate at the matched filter output from 4/Tc to 8/Tc, by means of digital interpolation to provide data at higher sampling rate to Rake finger.

The top-level architecture of DFE is given in Figure 2.

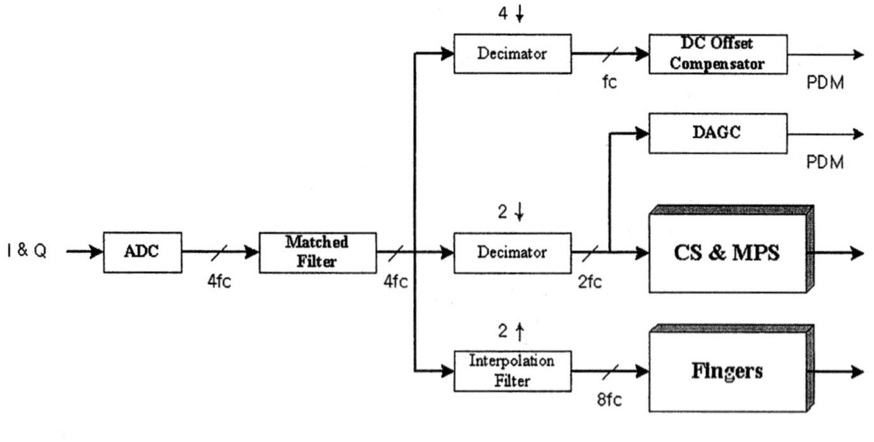

Figure 2. Top-level architecture of DFE.

3.2 Cell Searcher (CS)

In general, the CS, which results from the asynchronous behavior of the 3GPP network, searches over monitored and unlisted cells. Initial cell search, which is the task the UE has to perform first after power-on, is specifically related to the search of camping cell. In addition to the initial acquisition, the CS performs monitoring mode operation to monitor neighboring cells up to 32 by peak nulling method, which is necessary to identify possible candidates for either soft handover (SHO) or hard handover (HHO). The search for SHO candidate cells is performed on the same carrier frequency, which is currently used by the camping BTS (Basestation Transceiver System), while the search for HHO candidates is performed on different carrier frequencies upto 8. For that purpose, the camping BTS uses the feature of DL compressed mode (CM) to switch to that new target frequency during several slots of transmission gap.

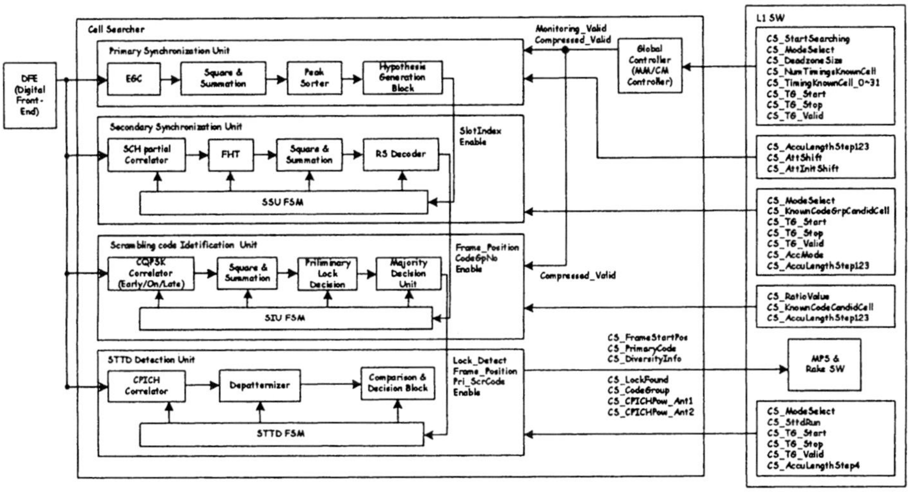

Figure 3. Top-level architecture of CS.

The acquisition procedure consists of three major steps and an additional step to detect the use of transmit diversity in the BTS. These four steps are

- 1st Step : Slot Timing Identification by PSC (Primary Synchronization Code) unit
- 2nd Step : Frame Timing & Scrambling Code Group Identification by SSC (Secondary Synchronization Code) unit
- 3rd Step : Scrambling Code Number Identification
- 4th Step : STTD Detection

CS global controller generates control signals out of a set of timing information given by the L1 SW. These control signals steer the acquisition procedure for all different operational modes of normal mode (NM), monitoring mode (MM) and compressed mode. The Top-level architecture of CS is given in Figure 3.

Each step has its own task, relies on intermediate results of the preceding step, and initiates the operation of the next step. All timing information that result from the cell searching operations or that are given to the CS by the SW as a priori knowledge have an accuracy of Tc/2. All absolute timing values are related to a UE global time reference: a counter that is running from 0 to $(15 \cdot 8 \cdot 2560-1)$ at 8/Tc. Since the accuracy of this counter is higher than the accuracy of the CS timing, a modified time based signal is fed into the CS.

The PSC unit has the task to process a configurable amount of input data and generate eventually a hypothesis on the timing of the slot boundaries in the received MF output signal, a value between 0 and $(2 \cdot 2560-1)$, together with an activation signal for the 2nd step. This PSC unit concerns all the operational modes of NM, MM and CM. Then, SSC unit copies the hypothesis for the slot boundary into an internal register and generates hypotheses on frame boundary timing, a value between 0 and $(15 \cdot 2 \cdot 2560-1)$, and on scrambling code group, a value between 0 and 63, together with an activation signal for the 3rd step. Then, the 3rd step copies these values to internal registers and tries to identify the used scrambling code inside the group reported by the 2nd step. If this operation ended successfully, the found absolute scrambling code number, a lock-found indicator, and the used timing information with an activation signal for the 4th step are generated. Then, the 4th step tries to detect whether STTD was used in the BTS transmitter and generates values that can be used to compute the CPICH power. The 4th step covers frequency error upto ± 3 ppm, i.e., ± 6 KHz. Denote CPICH is employed instead of SCH for STTD detection in this design.

The 2nd ~ 4th steps only concern the operational modes of NM and MM. The operations and communications described above result in one-way handshaking from the 1st to the 4th step. This control flow mechanism is more efficient than synchronizing the steps to a fixed timing grid that is spaced with the least common multiple timing of all units.

3.3 MultiPath Searcher (MPS)

The MPS establishes and updates the profile of the channel and selects echoes for further processing, i.e., tracking for demodulation, in the rake receiver based on the channel profile for 6 ~ 8 active set members. The MPS is composed of several functions that are described as follow:

- Correlator Bank
- Non-Coherent Accumulator
- Echo Selection Block

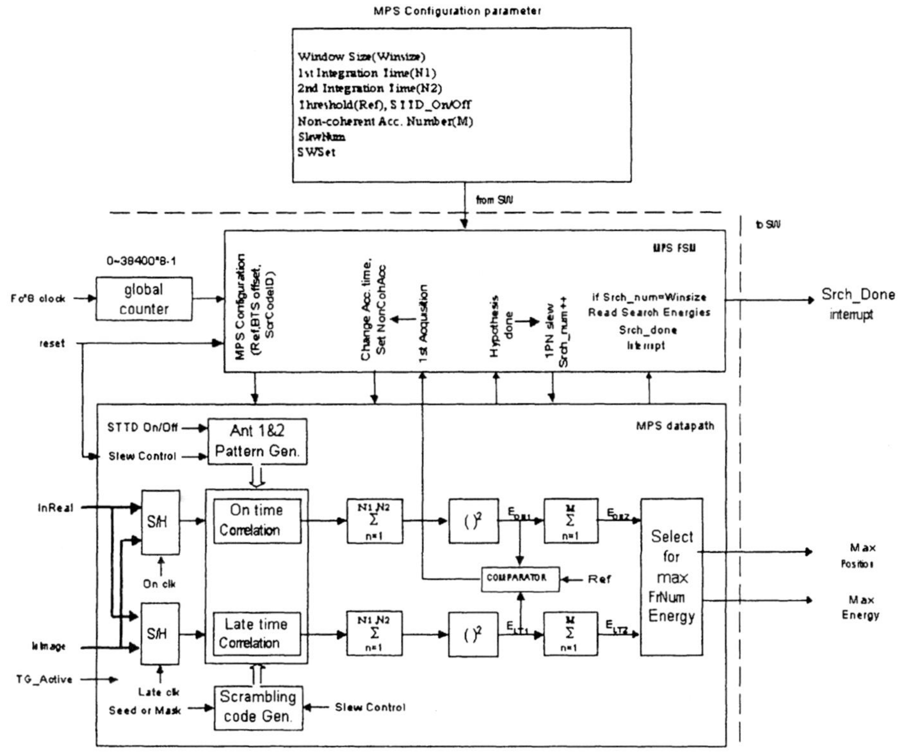

Figure 4. Top-level architecture of MPS.

The searcher engine searches strong echo signals in pair, evaluating the energy at both the current searcher position (On-Time) and a half PN chip delayed from the on-time position (Late-Time). Important performance criterion for MPS is searching time and its reliability. For fast searching the correlation length should be reduced as small as possible, and for reliability the correlation length should be increased as much as possible. For fast searching and high reliability simultaneously, we employ two different correlation lengths N_1 and N_2, where N_2 is greater than N_1. The resolution of N_1 and N_2 is Tc for each On-Time & Late-Time. If search energy for the correlation length N_1 is less than the threshold value given by SW, then searcher slews to the next position and starts correlation for this position. If search energy for the correlation length N_1 is greater than the threshold value, then searcher increases correlation length from N_1 to N_2 for that position. By using longer correlation length, detection probability can be increased and false alarm probability can be reduced. So, the double dwell serial searching algorithm can achieve fast searching time and reliability simultaneously. For more reliability, post-detection integration, i.e. non-coherent accumulation by M, is used additionally. That is, the total searching time per active member cells is the function of searching window size (L), correlation length (N_1, N_2), number of post-detection integration (M), number of multipaths and number of active cells. Figure 4 shows the top-level architecture of MPS.

The MPS consists of two major elements, the control finite state machine and the datapath. The SW configures the finite state machine by writing window size, N_1, N_2, M, threshold values, slew number and STTD_on, etc. The finite state machine forces the according generator set, consisting of scrambling code generator, OVSF code generator and pilot pattern generator, to slew to the correct search starting position. For search operation, the finite state machine looks for the next even CPICH symbol. If this occurs, the search over window size will be started. During this search, the finite state machine forces the other generator set to slew to the given search starting position. If one search is finished, the next search will be done with the other generator set. This dual generator set enhances the searching time performance, especially during SHO. Also, the finite state machine writes a SearchDone signal as a L1 SW interrupt. Due to this, the L1 SW reads the echo energy values and the according echo positions upto 4.

3.4 Rake Receiver and Combiner

Figure 5 shows the top-level architecture of Rake receiver and combiner. The L1 SW running on the u-Processor is responsible for finger assignment and finger control. The L1 SW gets information on the current finger state from the 4 Rake fingers and information on the available echoes in the transmission channels for all base stations in the active set from the MPS. The set of selected echoes from the MPS is determined by L1 SW. The information on echo power, base station ID and echo timing are updated by the interaction of L1 HW, L1 SW and high layer configuration.

There are 5 branches in the Rake receiver (per finger) and combiner.

- DPCH1 (DPCH1, DL-DPCCH, AICH, AP-ICH, CD/CA-ICH, PICH, CSICH)
- DPCH2 (DPCH2)

- DPCH3 (DPCH3, PCCPCH, PDSCH)
- SCCPCH (SCCPCH)
- CPICH for Rake or DPCH1Pilot for Combiner (dedicated DPCH1 branch for pilot symbols only)

The fingers perform channel estimation, physical channel demodulation, timing error detection and time tracking, and SIR and received signal code power (RSCP) estimation for finger lock detection. The feedback signaling message (FSM) encoder calculates the FSM information based on the channel estimations obtained from CPICH despreading for all fingers. The frequency error detector calculates estimates for the current frequency offset based on the channel estimations obtained from CPICH despreading for all fingers. The antenna verification unit is active only for closed-loop transmit diversity Mode 1. Antenna verification is performed based on the CPICH and DPCH based channel estimates in all active fingers. The antenna verification block feeds back the weights to be applied to the CPICH based channel estimation when applied to the demodulation of DPCH/PDSCH channels. UL and DL power control (PC) for DPCH are also the part of RAKE functionalities.

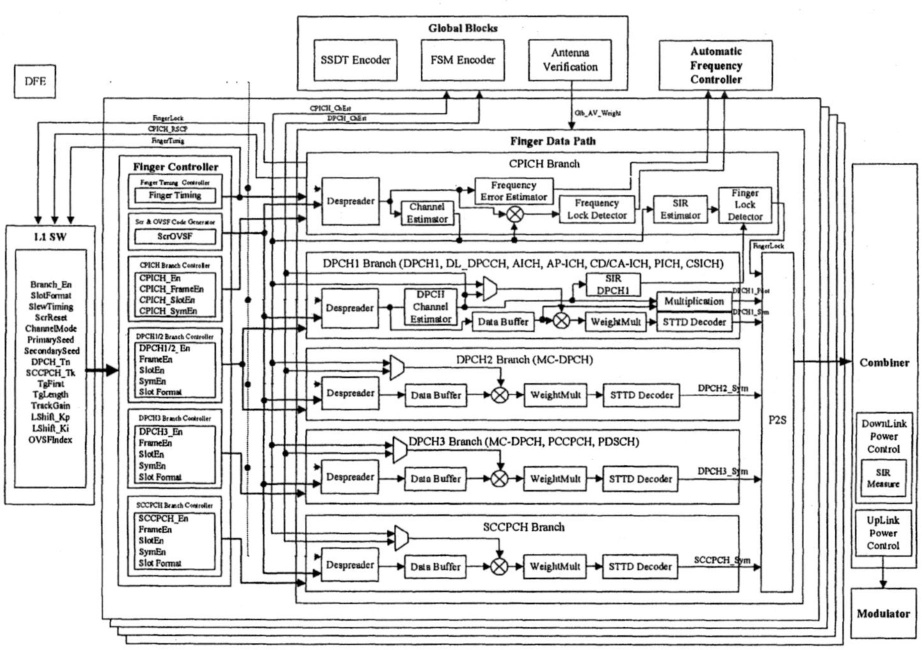

Figure 5. Top-level architecture of Rake receiver and combiner.

The combiner block performs coherent combining of the weighted and time-aligned demodulator output symbols, which are one type of user data with control information (e.g., DPCH) or pure control information (AICH, PICH). The combiner combines the received signal and demultiplexes user data and control information, i.e., user data to TrCH and TPC information to power control, etc. The combiner performs 3 tasks in the branches.

- Time Alignment of the Received Data in De-Skewing Buffer
- Combining of the Data in Data Combining Unit
- Decoding of Indicator Channels (AICH and PICH etc. in DPCH1 Branch only) in Indicator Channel Decoding Unit (in DPCH1 Branch Combining Unit)

For DPCH channels, this function is performed independent of the base station ID (i.e., for all active set members), while for all other channels, the combining is restricted to the set of fingers assigned to the echoes coming from one particular base station. The finger output symbols to the combiner are declared valid if the finger is enabled and the finger slewing procedures are finished, such that only useful symbols are actually taken into account for combining. In addition, the finger output symbols can be declared invalid if the finger is out-of-lock. This can be used to ensure a certain quality of the finger output symbols considered for combining.

4. UE MODULATOR

Figure 6 shows the top-level architecture of the modulator. The modulator supports upto 6 MCs, and it contains on-chip RRC filter. The modulator consists of the following functional blocks.

- Tx Controller
- Control Channel Generator
- OVSF Code Generator
- Scrambling Code Generator
- Spreading, Scrambling, Beta Multiplication, Channel Combining and Spreading
- Tx RRC Filter

The modulator has its interface with Demodulator, TrCH Encoder, RF and L1 SW of the modulator.

5. CONCLUSIONS

Design of 3GPP Compliant UE Modem, which has been implemented by top-down approach, has been described in this paper. Details of the Modulator and the Demodulator of DFE, CS, MPS, Rake receiver and Combiner are treated. The functions described in this paper have been tested successfully.

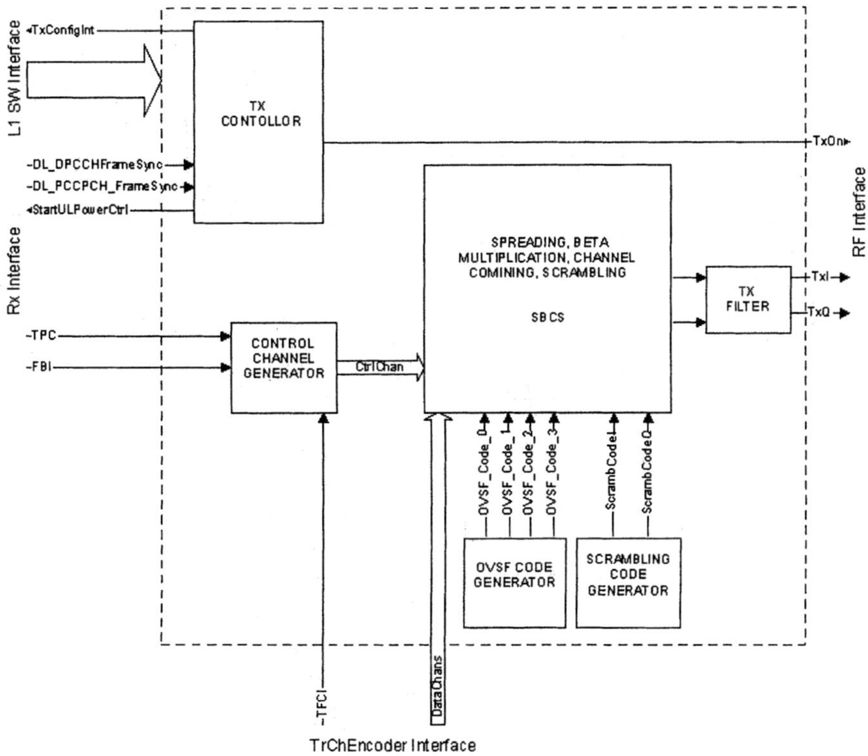

Figure 6. Top-level architecture of modulator.

ACKNOWLEDGEMENTS

The authors wish to thank the engineers at Synopsys Inc. for their sincere support on the project AIM (Asynchronous IMT-2000 Modem).

REFERENCES

[1] 3G TS 25.201 : "Physical Layer - General Description." Release 1999.

[2] 3G TS 25.211 : "Physical Channels and Mapping of Transport Channels onto Physical Channels (FDD)." Release 1999.

[3] 3G TS 25.212 : "Multiplexing and Channel Coding (FDD)." Release 1999.

[4] 3G TS 25.213 : "Spreading and Modulation (FDD)." Release 1999.

[5] 3G TS 25.214 : "Physical Layer Procedures (FDD)." Release 1999.

[6] 3G TS 25.215 : "Physical Layer - Measurements (FDD)." Release 1999.

[7] 3G TS 25.101 : "UE Radio Transmission and Reception (FDD)." Release 1999.

[8] 3G TS 34.121 : "Terminal Conformance Specification; Radio Transmission and Reception (FDD)." Release 1999.

Chapter 6

MMSE PERFORMANCE OF ADAPTIVE OVERSAMPLING LINEAR MULTIUSER RECEIVERS IN CDMA SYSTEMS

Phichet Iamsa-ard and Predrag B. Rapajic
The University of New South Wales, Sydney, NSW 2052, Australia

Abstract: An adaptive linear MMSE receiver for CDMA systems is introduced; subsequently, the conjecture about the avoidance of oversampling is addressed. Simulation results highlight a trade-off between performance degradation during application of conventional *chip-duration-spaced* sampling with selected signature sequences and an increased cost of complexity for oversampling implementation. In additional, the suggestion on the signature sequence design for CDMA systems using the chip-duration-spaced sampling with the adaptive linear MMSE receiver is also given.

Key words: Adaptive filters, Code division multi-access, FIR digital filters, Multiuser channels.

1. INTRODUCTION

Code-division multiple-access (CDMA) is a multiplexing technique in which several independent users access a common communications channel by modulating their information input with their unique signature sequences. A lot of receiver structures have been investigated [1]. One receiver structure "adaptive linear MMSE receiver" proposed in [3] has been shown to have several advantages over the conventional matched filter based receiver [4]. Its structure is attractive because it can be implemented in a decentralized fashion. Also, it can eliminate multiple access interference, as the centralized multi-user detectors while disregarding information on timing, signature, carrier phase and initial synchronization are needed as the conventional receiver.

However, the implementation of the adaptive linear MMSE receiver is originally realized on the fractionally-spaced transversal filters [3] which received signals are oversampled by an input sampler in order to moderate the effect of timing phase error. The idea of oversampling received signals is adapted from the concept of the fractionally spaced equalizer [7-8]. By the realization of oversampling, memory required for receiver coefficients and calculation complexity is increased. Also, additional algorithms such as the tap leakage algorithm [9] for the stable operation of this receiver may be needed.

In this chapter, it is shown that the necessity for oversampling in an adaptive linear MMSE receiver may be dropped or alleviated by appropriate selection of signature sequences. Moreover, the sensitivity to a timing phase error of signature sequences is also investigated.

2. SYSTEM DESCRIPTION

We consider a discrete-time, baseband K-user CDMA system as depicted in Figure 1a. We further consider an information symbol of the k-th user $a_k(i) \in \{-1,1\}$ having a period T modulated by an impulse signature sequence assigned to the user,

$$s_k(t) = \sum_{m=0}^{M-1} c_k(m)\delta(t - mT_c) \tag{1}$$

where T_c is the chip duration referring to the interval during any two adjoining signature symbols, $c_k(m)$ and $c_k(m+1)$. The spreading gain for this signature is thus 1/M. Subsequently, the modulated-impulse signal is filtered by the waveshaping filter $p(t)$ to limit the transmission bandwidth.

At the receiving end shown in Figure 1b, the noise at the sampler input. The received signal of the K-user asynchronous system at the adaptive MMSE filter may be expressed as

$$y(t) = \sum_{k=1}^{K}\sum_{i=0}^{N-1} a_k(i)u_k(t - iT - \phi_k) + v(t) \tag{2}$$

where $u_k(t) = s_k(t) \otimes p(t) \otimes h(t,\tau)$ refers to the received signature waveform as seen by the adaptive MMSE receiver. ϕ_k is an arbitrary reference to the k-th user; $\phi_k \in [0,T)$. $h(t,\tau)$ is the slow-fading wide-sense-stationary scattering (WS-SUS) channel and $v(t)$ is bandlimited additive white Gaussian noise (AWGN) with variance σ^2. The received signal is sampled at rate $1/T_s$ where $T_s = DT_c$ and D is integer. The discrete T_s-spaced sampled output is fed to the adaptive MMSE filter afterwards.

After the adaptive filtering process, a downsampler of the symbol rate T is applied to the adaptive linear filter output, implying a cascaded adaptive linear filter and output sampler serving as a decimating filter.

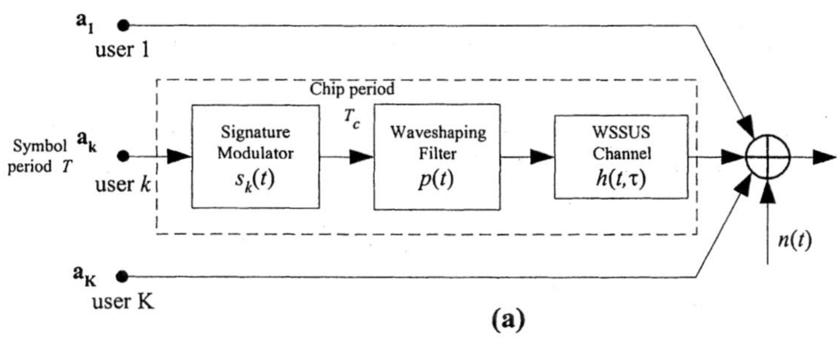

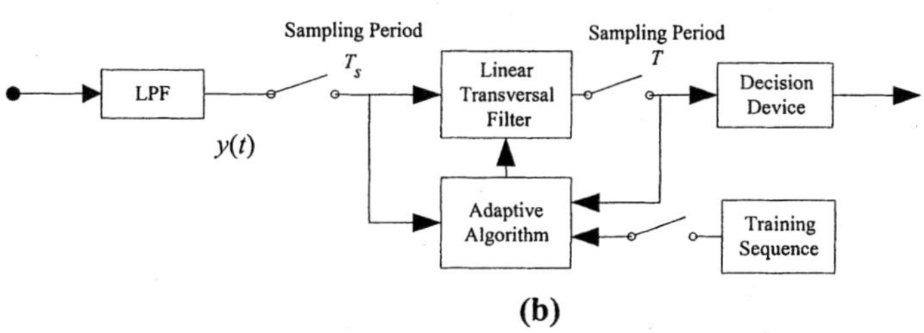

Figure 1. (a) Transmission of K-user CDMA system
(b) Adaptive MMSE receiver for the desired k-user

3. THE ADAPTIVE MMSE RECEIVER

3.1 The Optimum MMSE Filter Coefficients

Basically, the adaptive MMSE filter can be realized in the form of linear transversal filter. By this, the sampled signal will be observed over a running window of length L; the observed window can be represented in matrix form as follows:

$$\mathbf{y}(i) = \mathbf{Ua}(i) + \boldsymbol{\nu}(i) \tag{3}$$

where $\mathbf{U}$ is the matrix of the received sampled signature waveform; $\mathbf{U} = [\mathbf{u}_1, \mathbf{u}_2, \ldots, \mathbf{u}_K]$, the vector $\mathbf{a}(i) = [\mathbf{a}_1, \mathbf{a}_2, \ldots, \mathbf{a}_K]$ represents transmitted symbols of K users during that window period and $\boldsymbol{\nu}(i)$ is bandlimited AWGN.

By applying the mean-squared error (MSE) criterion [2], the expression of the optimum linear FIR filter coefficients for the desired k-th user is given by [3];

$$\mathbf{w}_{\text{opt}} = \mathbf{F}^{-1}\mathbf{u}_{\mathbf{k}} \tag{4}$$

where the matrix $\mathbf{F} = \mathbf{UU}^{\mathrm{H}} + \boldsymbol{\sigma}^2$ is nonsingular [5] and called the *multiple access channel correlation matrix*, and $\mathbf{u}_{\mathbf{k}} = E[a_k \mathbf{y}]$ is the signature vector associated with symbol a_k.

For conventional T_c-spaced sampling or sampling with period $T_s = T_c$, the transfer characteristic of this optimum linear FIR filter was derived in the Appendix A of [3] shown as

$$W_k(\omega) = \frac{u_k^*(\omega)}{\frac{1}{T_c}\sum_{k=1}^{K}\sum_{l=0}^{L-1}\left|u_k(\omega + l\frac{2\pi}{T_c})\right|^2 + \sigma^2} \tag{5}$$

where $u_k(\omega)$ is the Fourier transform of the received signature waveform of the k-th user and the number of filter coefficients is L-1. Note that $u_k(\omega)$ in the denominator of (5), is the aliased spectrum of sampled signature waveform. However, to consider solely on the spectrum of the interest user; user k, we may write (5) as follows

$$W_k(\omega) = \frac{u_k^*(\omega)}{\frac{1}{T_c}\sum_{l=0}^{L-1}\left|u_k(\omega + l\frac{2\pi}{T_c})\right|^2 + \Psi + \sigma^2} \tag{6}$$

where Ψ denotes the *multiple access interference* (MAI).

Moreover, it is assumed that the coefficients spanning LT_c of the linear FIR filter is sufficiently long to collect most of the energy dispersed by the transmission channel and the waveshaping filter.

3.2 Effects of the Timing Phase Error

Let the k-th user be the user of interest. For simplicity of notation, we may omit the subscript for user k in the sequel. Referring to the model in Figure 1b, the aliased (Nyquist equivalent) spectrum (noise ignored) of an isolated chip observed by the adaptive MMSE filter is thus

$$\mathcal{U}_{\mathrm{eq},\tau}(\omega)=\sum_{l}\mathcal{U}_{k}(\omega+l\frac{2\pi}{T_c})\exp\{j(\omega+l\frac{2\pi}{T_c})\tau\} \tag{7}$$

where the sampling instants of the input sampler are at $t=iT_c+\tau$. We further assume that if the *excess bandwidth* occupied by the signal beyond the Nyquist frequency caused by the waveshaping filter $p(t)$ is less than 50%, it is then easy to show that only three terms corresponding to $l=0,\pm 1$ will survive. We may thus express the square magnitude of as $\mathcal{U}_{\mathrm{eq},\tau}(\omega)$

$$\left|\mathcal{U}_{\mathrm{eq},\tau}(\omega)\right|^2=\left|\mathcal{U}_{k}(\omega)e^{j\omega\tau}+\mathcal{U}_{k}(\omega-\frac{2\pi}{T_c})e^{j(\omega-\frac{2\pi}{T_c})\tau}\right|^2 ;\left|\omega\right|\le\frac{\pi}{T_c} \tag{8}$$

There are some observations about $\left|\mathcal{U}_{\mathrm{eq},\tau}(\omega)\right|^2$ in (8). First, a null may be created in the rollout portion $(1-\alpha)\pi/T_c\le|\omega|\le(1+\alpha)\pi/T_c$; that is, the width of this null is determined by the rolloff factor α while its depth depends on τ which gives the deepest null when $\tau=T_c/2$. Note that the null will cause performance degradation because of the noise and MAI enhancement resulted by an attempt of the linear FIR filter to compensate for this null.

Secondly, since $\mathcal{U}_{\mathrm{eq},\tau}(\omega)$ is the cascade of the signature modulator, waveshaping filter and transmission channel, the characteristic of each signature sequence may result in the different severity level of the null. This is discussed further in the next session.

To mitigate the effects of noise and MAI enhancement, an oversampling technique may be applied to the input sampler. This will enable the adaptive MMSE filter to compensate for the channel distortion directly, before aliasing and later compensating for any arbitrary timing phase distortion [7-9]. Unfortunately, sampling at higher rate requires more memory for coefficients and complex coefficient calculations are required to keep the same time span of the linear FIR filter; these may be undesired in some circumstances.

3.3 The Influence of Signature Sequences

Recalling (1), we may write the transfer function of the signature modulator as follows

$$S(\omega) = c_0 + c_1 e^{-j\omega T_c} + \ldots + c_{M-1} e^{-j(M-1)\omega T_c} \tag{9}$$

For the waveshaping filter, we make our consideration practical by applying the raised cosine frequency characteristic:

$$P(\omega) = \begin{cases} T_c & ;0 \le |\omega| \le (1-\alpha)\dfrac{\pi}{T_c} \\ \dfrac{T_c}{2}\left\{1+\cos\left[\dfrac{T_c}{2\alpha}\left(|\omega|-(1-\alpha)\dfrac{\pi}{T_c}\right)\right]\right\}; & \\ \qquad (1-\alpha)\dfrac{\pi}{T_c} \le |\omega| \le (1+\alpha)\dfrac{\pi}{T_c} & \\ 0 & ;|\omega| \ge (1+\alpha)\dfrac{\pi}{T_c} \end{cases} \tag{10}$$

where α is the rolloff factor. Let $\boldsymbol{H}(\omega)$ be the transfer function of the cascade of $\boldsymbol{S}(\omega)$ and $\boldsymbol{P}(\omega)$,

$$\boldsymbol{H}(\omega) = \boldsymbol{S}(\omega)\boldsymbol{P}(\omega) \tag{11}$$

We show the frequency characteristics $|\boldsymbol{H}(\omega)|^2$ of several signature sequences when their sampled phases are 0 and $T_c/2$, and the rolloff factor of the waveshaping filter is equal to 0.5 in Figure 2. It is shown that the energy distribution of each sequence depends on the signature sequence. Furthermore, when the received signal is sampled at $\tau = T_c/2$, it is seen that the more energy in the affected region $(1-\alpha)\pi/T_c \le |\omega| \le (1+\alpha)\pi/T_c$, the more energy will be eliminated from this frequency range. For instance the energy of the sequences in Figure 2a and Figure 2c are concentrated near π/T_c, thus when the received signal is sampled at $\tau = T_c/2$, most of spectrum energy remains. On the other hand, the sequences in Figure 2b and *Figure 2d* have the energy distribution mainly in the band $(1-\alpha)\pi/T_c \le |\omega| \le (1+\alpha)\pi/T_c$; therefore, when the null is created at this region, the majority of their energy will be wiped out.

Since the linear FIR filter will make an inversion of channel response plus MAI and noise, this leads MAI and noise enhancement in that affect region. However, the severity o the enhancement depends on the energy wiped out by the aliasing effects; that is, the signature sequence may be another key role to alleviation enhancement problems without applying oversampling. For instance, the signature sequences in *Figure 2b* and *Figure 2d* which are very sensitive to the timing phase error can cause the higher enhancement of noise and MAI than applying the sequences in *Figure 2a* and *Figure 2c* when the timing phase error occurs.

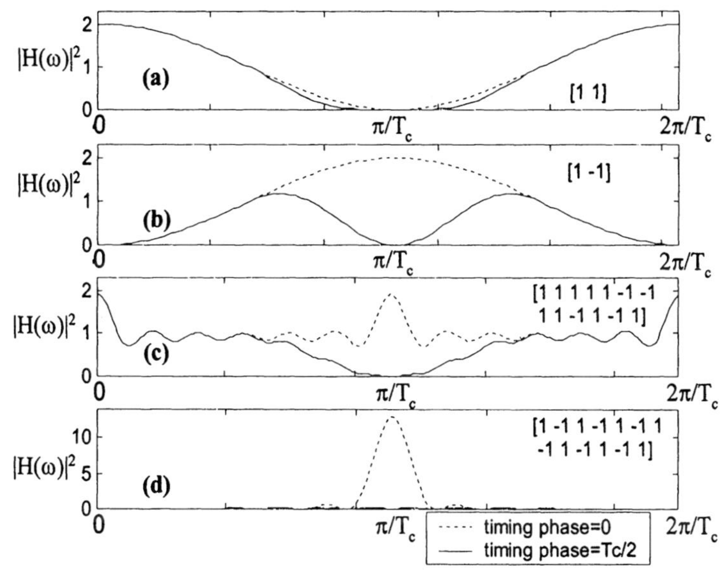

Figure 2. Plot of $|\boldsymbol{H}(\omega)|^2$ in the case $\tau = 0$ and $T_c/2$ the sequence
(a) [1 1] (b) [1 -1] (c) [1 1 1 1 1 -1 -1 1 1 -1 1 -1 1]
(d) [1 -1 1 -1 1 -1 1 -1 1 -1 1 -1 1]
Note that the rolloff factor α=0.5 for all plots.

Another point to be noted is that the longer the sequences, the greater the choices (degree of freedom) when selecting the proper signature sequences.

Now the problem of oversampling avoidance is narrowed to how many *good* sequences can be assigned and how much of a trade-off between the cost of complexity for oversampling and the performance degradation for conventional sampling. We will address these through simulation results in the next section.

4. NUMERICAL RESULTS

CDMA systems with a spreading gain 8 and 12 are considered in Figure 3 and Figure 4, respectively. The performance index, mean-squared error (MSE), is computed at the steady-state in the training process where the least-mean-square (LMS) algorithm is applied. Furthermore, the performance index is observed when the timing phase is at $\tau = 0$, $T_c/4$ and $T_c/2$ of T_c-spaced sampling. The timing phase $\tau = 0$ is intentionally chosen to show the perfect case while $\tau = T_c/2$ is for illustrating the worst case of sampling. In addition, the MSE of the adaptive MMSE receiver applying oversampling ($T_c/2$-spaced) is also given.

We assume herein that the structure of MAI is ignored and the MAI is assumed to be Gaussian distributed. The signal-to-noise-plus-MAI (SNIR) ratio is set to 20 dB. Moreover, the waveshaping filter having a raised cosine spectrum with the rolloff factor α=0.5 is applied. The tails of its impulse response are limited at magnitude of 15 dB below its peak value. Note that although the different choices of the rolloff factor value result in the MSE, by simulation experience results when the rolloff factor in the range 0.1-1.0 are not far different; therefore, the only selected value of α=0.5 is sufficient to draw a conclusion.

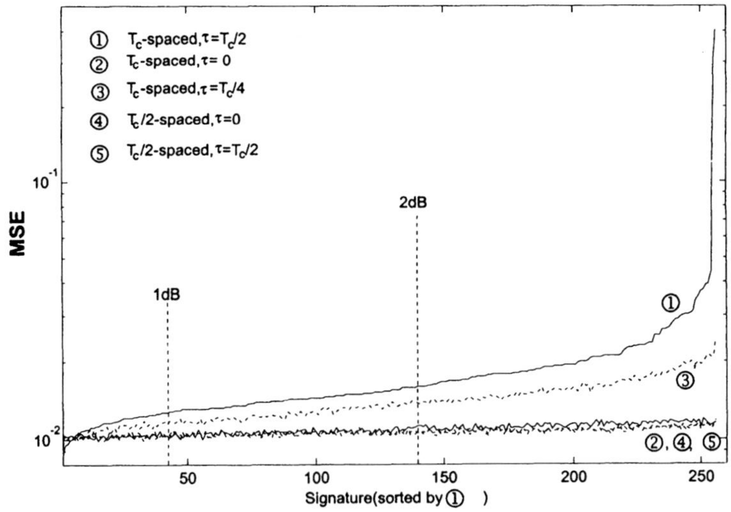

Figure 3. Plot of MSE versus signature sequences (sorted) of length 8 and SNIR=20dB

In Figure 3 and Figure 4, the MSE is plotted in logarithm scale against all possibilities of signature sequences, which can be represented from 0 to 255 and 4095 (decimal) for signature sequence of length 8 and 12, respectively. To express the influence of signature sequences explicitly, we have sorted the MSE in ascending order of the T_c-spaced, $\tau = T_c/2$ category so as to present the relation of the number of signature sequences (in worst case of sampling) effecting to each level of the MSE. Then the MSE of the other categories are plotted with respect to the reordered signature sequences. As expected, all sequences of the conventional T_c-spaced sampling, perfect sampling phase $\tau = 0$ (②), the $T_c/2$-spaced sampling, perfect sampling phase $\tau = 0$ (④), and the $T_c/2$-spaced sampling, sampling phase $\tau = T_c/2$ (⑤) have the same level of MSE while each sequence in the conventional T_c-spaced sampling where sampling phases are not perfect; $\tau = T_c/2$ (①) and $T_c/4$ (③) has different level of MSE starting from the same level as in the perfect sampling phase category to very high MSE.

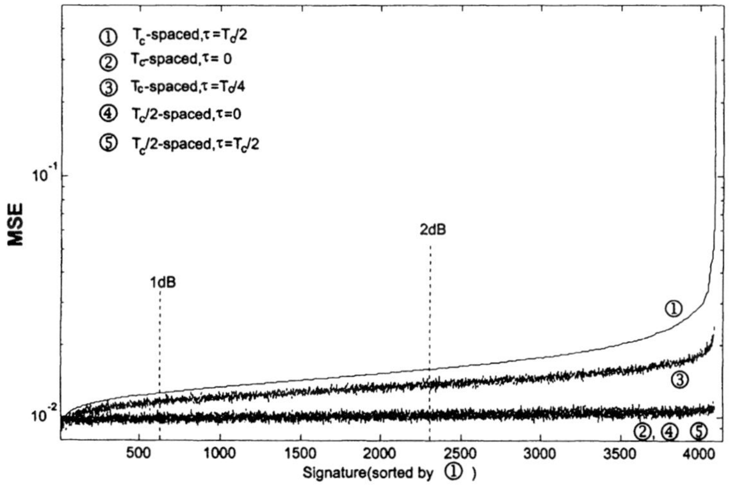

Figure 4. Plot of MSE versus signature sequences (sorted) of length 12 and SNIR=20dB

From the figures, we can draw a conclusion that:

- Every sequence yields the same level of MSE when the timing phase of the conventional T_c-spaced sampling or the $T_c/2$-spaced sampling applied.
- When the conventional T_c-spaced sampling applied with imperfect sampling phase, each sequence has the MSE response differently varying with the sampling phase τ.

From the simulation results, the problems arising from the previous section about the trade-off between the conventional sampling and oversampling, and how many good sequences existing can be answered. We note that the worst category; T_c-spaced, $\tau = T_c/2$, will be used for the following comparison.

In Figure 3, there are approximately 42 (16%) signature sequences can be used for T_c-spaced sampling indicating the MSE degradation within 1 dB of those oversampling whereas we can apply 150 (55%) signature sequences for T_c-spaced sampling yielding the MSE degradation worse within 3 dB of those oversampling.

Similarly, in Figure 4 there are about 625 (15%) and 2,300 (56%) signature sequences can be used for T_c-spaced sampling giving the MSE degradation within 1 dB and 2 dB, respectively.

We also note that the sequences showing high sensitiveness to the timing phase error have their energy distribution mainly in $(1-\alpha)\pi/T_c \leq |\omega| \leq (1+\alpha)\pi/T_c$ region. Not surprisingly, the two signature sequences giving the largest MSE for T_c-spaced sampling, when $\tau = T_c/2$ and $T_c/4$ showing in Figure 3 and Figure 4 are the sequences alternating between 1 and -1; [1 -1 1 -1 1 -1 ...] and [-1 1 -1 1 -1 1 ...] of length 8 and 12, respectively.

Since the simulation results show the degree of timing phase error sensitiveness of each sequence where some sequences are acutely sensitive to the timing phase error; therefore, the avoidance of these sequences for conventional T_c-spaced sampling can prevent the system from vicious degradation of performance.

As described above, the Figure 3 and Figure 4 explicitly show the trade-off between complexity increased from implementing the oversampling and the performance degradation from applying the conventional T_c-spaced sampling.

5. CONCLUSIONS

The avoidance of oversampling for adaptive linear MMSE receivers for CDMA systems (Gaussian-distributed MAI assumed) is addressed.

The simulation results have shown that by applying the conventional T_c-spaced sampling, each signature sequence responds differently to the timing phase error with different value of mean-squared error (MSE). Some sequences are negligible sensitive and some are highly sensitive to the timing phase error.

It was illustrated that the choices of spreading sequences affect the performance when the conventional T_c-spaced sampling applied with imperfect timing phase. To alleviate the timing phase problem, an oversampling of received signal with proper sampling rate was applied [7] and widely accepted. However, although the oversampling can effectively eliminate the sensitivity of timing phase of signature sequences, it costs complexity and calculation increased.

Alternatively, we have showed that by selection of the signature sequences that are less responsive to the timing phase error with an acceptable level of performance degradation, we can have a reasonable trade-off between the cost of complexity of oversampling implementation and the performance degradation of the conventional T_c-spaced sampling. In other words, each signature sequence is different in sensitiveness to the timing phase error when chip-rate sampling is applied. The less sensitive sequences yield the MSE close to the oversampling case or chip-rate sampling with perfect timing phase.

Although the results of the degradation level shown in Figure 3 and Figure 4 are the special case when the rolloff factor value $\alpha = 0.5$ and varying the rolloff factor α may change the MSE, the same conclusion can still be drawn as above.

In addition, we have observed that the desired signature sequences should have spectrum energy concentrated as less as possible in the rolloff region; $(1-\alpha)\pi/T_c \le |\omega| \le (1+\alpha)\pi/T_c$ where α is the rolloff factor. This can be a criterion for the signature sequence design for CDMA systems applying adaptive linear MMSE receivers where the T_c-spaced sampling is employed.

REFERENCES

[1] S. Verdu, *Multiuser Detection.* Cambridge, U.K.: Cambridge Univ. Press, 1998.

[2] J. G. Proakis, *Digital Communications.* New York: Wiley, 1989.

[3] P. B. Rapajic and B. S. Vucetic, "Adaptive receiver structures for asynchronous CDMA systems," *IEEE J. Select. Areas Commun.*, vol. 12, pp. 685-697, May 1994.

[4] D. K. Borah and P. B. Rapajic, "Adaptive MMSE Maximum Likelihood CDMA Multiuser Detection," *IEEE J. Select. Areas Commun.*, vol. 17, pp. 2110-2112, Dec. 1999.

[5] F. Ling, S. U. H. Qureshi, "Convergence and Steady-State Behavior of a Phase-Splitting Fractionally Spaced Equalizer," *IEEE Trans. Commun.*, vol. 38, pp. 418-425, Apr. 1990.

[6] P. Iamsa-ard, *Masters Thesis.* UNSW, 2001.

[7] R. D. Gitlin and S. B. Weinstein, "Fractionally-spaced equalization: an improved digital transversal equalizer," *Bell Syst. Tech. J.*, vol. 60, no. 2, pp. 275-296, Feb. 1981.

[8] G. Ungerboeck, "fractional tap-spacing equalizer and consequences for clock recovery in data modems," *IEEE Trans. Commun.*, vol. COM-24, no. 8, pp. 856-864, Aug. 1976.

[9] R. D. Gitlin, H. C. Meadors, Jr., and S. B. Weinstein, "The Tap-Leakage Algorithm: An Algorithm for the Stable Operation of a Digitally Implemented, Fractionally Spaced Adaptive Equalizer," *Bell Syst. Tech. J.*, vol. 61, no. 8, pp. 1817-1839, Oct. 1982.

Chapter 7

PEAK-TO-AVERAGE POWER RATIO OF IEEE 802.11A PHY LAYER SIGNALS

A. D. S. Jayalath and C. Tellambura
School of Computer Science and Software Engineering, Monash Univeristy, Clayton, VIC 3800, Australia

Abstract: In this chapter, we propose clipping with amplitude and phase corrections to reduce the peak-to-average power ratio (PAR) of orthogonal frequency division multiplexed (OFDM) signals in high-speed wireless local area networks defined in IEEE 802.11a physical layer. The proposed technique can be implemented with a small modification at the transmitter and the receiver remains standard compliant. PAR reduction as much as 4dB can be achieved by selecting a suitable clipping ratio and a correction factor depending on the constellation used. Out of band noise (OBN) is also reduced.

Key words: OFDM, Peak-to-average power ratio, IEEE 802.11a PHY.

1. INTRODUCTION

With a rapidly growing demand for wireless communications, much research has been expended on providing efficient and reliable high-data-rate wireless services. The IEEE 802.11 standard for wireless local area networks (WLAN) was first established in 1997, and it supported data rates of 1 Mb/s and 2 Mb/s in indoor wireless environments. In comparison to 100 Mb/s Ethernet, the 2 Mb/s data rate is relatively slow and is not sufficient for most multimedia applications. Recently, the IEEE 802.11 WLAN standard group finalized the IEEE Standard 802.11a, which has an OFDM physical layer for indoor wireless data communications [1]. The data rates of IEEE 802.11a range from 6 Mb/s up to 54 Mb/s. This new standard can provide almost all multimedia communication services in indoor wireless environments.

However, two limitations of OFDM based systems are often asserted. Firstly, due to the nonlinearities of transmitter power amplifiers, high PAR values of the OFDM signal generate high OBN. There are restrictions imposed by the Federal Communications Commission and other regulatory bodies on the level of these spurious transmissions. These restrictions impose a maximum output power limitation, which determines power amplifier back-off. Further more nonlinearities of amplifier cause inband distortion of the signal giving higher bit error rates (BER). 64-quadrature amplitude modulation (64-QAM) modulation is used in IEEE.11a PHY for 48 Mb/s and 54 Mb/s transmission. This modulation scheme is highly sensitive to distortions due small Euclidian distances between signal points. Thus IEEE 802.11a WLAN devices may need power amplifiers with large back off, which are inefficient and bulky. Overcoming this problem requires reducing the PAR of OFDM signals. Secondly, OFDM is highly sensitive to frequency off-set errors. However, frequency offset due to the mobility of the terminal is negligible in WLAN environments, owing to very low speeds of the mobile.

Clipping high peaks increases the in-band distortion and the OBN and several studies of clipped OFDM signal have been reported. In [2] several modulation schemes are examined and effects of amplitude limiting are presented. A controlled amount of limiting is permissible in many cases. Clipping an oversampled signal produces less in-band noise but the OBN will increase [3]. Performance evaluation of clipped OFDM symbol with and without oversampling is presented in [4]. Two extra oversampled DFT operation are used in the transmitter to filter the OBN due to clipping. Oversampling is necessary to avoid peak regrowth after filtering. Similar PAR reduction scheme based on clipping and filtering an oversampled signal is presented in [5]. Two oversampled inverse fast Fourier transform (IFFT) operations are used to interpolate and perform frequency domain filtering after clipping. This reduces the amount of OBN. Increased BER due to clipping may be reduced using forward error correction coding. Thus, a system based on clipping and forward error correction is proposed in [6]. Decision aided reconstruction (DAR) is proposed in [7] for mitigating the clipping noise. The receiver is assumed to know the clipping level. DAR is an iterative reconstruction technique, which increases the receiver complexity. Several discrete Fourier transform (DFT) operations are performed at the receiver before reconstructing the original signal.

For baseband transmission, an approximation for the resulting increase in BER is given in [8], but spectral distribution of the distortion is not considered. References [9, 10] consider the spectral spreading but only for real valued discrete multitone (DMT) signals. The degradation due to clipping of an OFDM signal is analyzed in [11]. Extensive research has been undertaken to understand the effect of high power amplifier (HPA) nonlinear distortion in OFDM systems [12, 13]. Compared to other PAR reduction schemes clipping based techniques are simple to implement.

A PAR reduction using signal clipping and phase correction at the transmitter is proposed in [14]. The basic idea is as follows. Let the IFFT of the modulated symbol sequence $X_n, n = 0,1,\ldots N-1$ be $x_k, k = 0,1,\ldots N-1$. Here we assume quadrature phase shift keying (QPSK). The sequence x_k is clipped to a desired level, and the IDFT is taken giving out $\hat{X}_n$. Because of the clipping, we have $\hat{X}_n \neq X_n$. The key idea in [14] is to apply phase correction to $\hat{X}_n$ such that both $\hat{X}_n$ and X_n have identical phases. This is done by multiplying $\hat{X}_n$ with $e^{j(\theta_n - \hat{\theta}_n)}$ where θ_n and $\hat{\theta}_n$ are the phases of X_n and $\hat{X}_n$. The idea of this phase correction is to limit OBN and BER degradation. Of course, we cannot make the amplitude $\hat{X}_n$ equal to the original signal magnitude. Then we would be back to the original input sequence. Finally, the IDFT of the phase-corrected sequence $\hat{X}_n$ is used to generate the transmitted signal. Note that only N samples per each OFDM symbols are generated. Clipping those to a low level does not guarantee that the peaks of the output signals are reduced to the same level.

This chapter presents three enhancements to the method proposed in [14]. First, we oversample the OFDM signal by a factor of L. Oversampling will lead to greater robustness against peak regrowth. Oversampling is implemented using LN-length IDFTs. Secondly, the phase correction is controlled (additional parameter β) and is applied only if the difference between θ_n and $\hat{\theta}_n$ exceeds β. Third, if the amplitude deviates too much from the original signal amplitude (as measure by a parameter α), an amplitude adjustment is applied. The BER degradation caused is thus negligible and the OBN is reduced considerably. This approach is more general than [14]. For instance the special case in our scheme represented by $\alpha \to \infty$ and $\beta = 0$ corresponds to [14]. We then apply this new scheme in IEEE 802.11a PHY transmit signals and evaluate the performance.

2. AN OFDM SYSTEM AND PEAK-TO-AVERAGE POWER RATIO (PAR)

A block of N symbols, $\mathbf{X} = \{X_1, X_2, \ldots, X_{N-1}\}$, is formed with each symbol modulating one of a set of N subcarriers with frequency, $f_n, n-0,1,\ldots,N-1$. The N subcarriers are chosen to be orthogonal that is

$f_n = n\Delta f = n/T$, where and T is the OFDM symbol duration. The complex baseband signal can be expressed as

$$x(t) = \frac{1}{\sqrt{N}} \sum_{n=0}^{N-1} X_n e^{j2\pi f_n t}, \quad 0 \le t \le T \tag{1}$$

This signal can be generated by taking an N point inverse discrete Fourier transform (IDFT) of the block $\mathbf{X}$ followed by low pass filtering. The actual transmitted signal is modeled as $\text{real}\left[x(t)e^{j2\pi f_c t}\right]$, where f_c is the carrier frequency. The PAR of the transmitted signal in (1) can be defined as

$$\xi = \frac{\max |x(t)|^2}{E[|x(t)|^2]} \tag{2}$$

where $E[x]$ is the expected value of x. The PAR of the continuous-time OFDM signal cannot be computed precisely by the use of the Nyquist sampling rate [15], which amounts to N samples per symbol. In this case, signal peaks are missed and PAR reduction estimates are unduly optimistic. Oversampling by a factor of 4 is sufficiently accurate and is achieved by computing the $4N$ -point IDFT of the data frame.

3. IEEE 802.11 *a* SYSTEM DESCRIPTION

The IEEE 802.11 transmitter with the proposed PAR reduction scheme is presented in Figure 1. Input data is first mapped into N symbols (X_n) and serial to parallel converted. Then a LN point oversampled IFFT is taken. The output of oversampled IFFT is clipped according to the clipping ratio selected. A second LN point fast Fourier transform (FFT) is taken to get the clipped sample points back to frequency domain. We will now select N samples corresponding to the original signals (X_n). The clipping of the signal causes dispersion of the signal points. If we do not clip the signal, the original symbols (X_n) will be regenerated at this point.

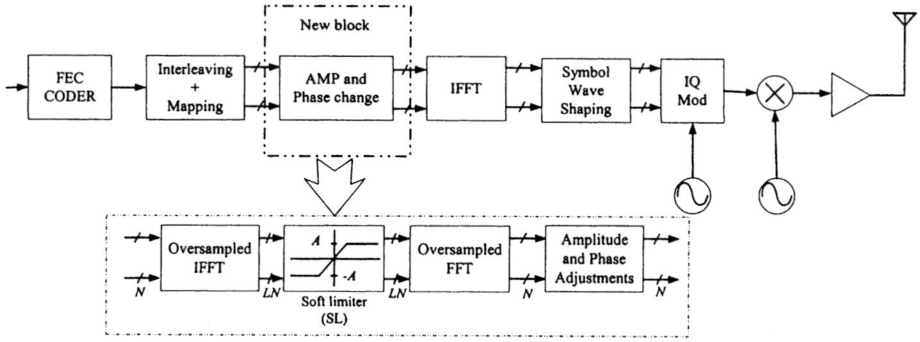

Figure 1. IEEE 802.11a Transmitter

Next, we adjust the phase and the amplitude of these signal points $\hat{X}_n$ such that they are confined to a smaller region around the original signal points X_n.

Let us define an arbitrary signal point of the given constellation having amplitude and phase γ_n and θ_n as shown in Figure 2. Then phase, $\hat{\theta}_n$ and amplitude $\hat{\gamma}_n$ of each signal point $\hat{X}_n$ is adjusted according to the following rules. If the amplitude of $\hat{X}_n$ deviates from γ_n by more than α, the amplitude will be adjusted as follows.

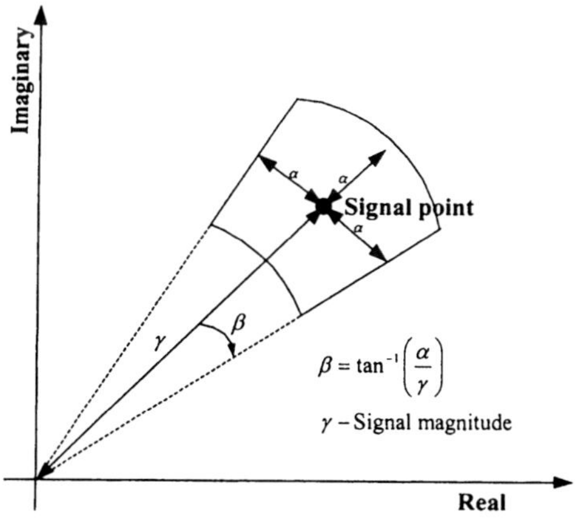

Figure 2. Amplitude and phase adjustment parameters

$$\hat{\gamma}_n = \begin{cases} \gamma_n + \alpha & \hat{\gamma}_n - \gamma_n > \alpha \\ \gamma_n - \alpha & \gamma_n - \hat{\gamma}_n > \alpha \\ \gamma_n & \text{else} \end{cases} \quad n = 0,1,\ldots N-1 \tag{3}$$

If the phase of the clipped signal deviates by β or more from the phase of the original signal points, the phase is also adjusted. β is calculated from the given α as follows

$$\beta = \tan^{-1}\left(\frac{\alpha}{\gamma_n}\right) \tag{4}$$

$$\hat{\theta}_n = \begin{cases} \theta_n + \beta & \hat{\theta}_n - \theta_n > \beta, \\ \theta_n - \beta & \theta_n - \hat{\theta}_n > \beta, \quad n = 0,1,\ldots N-1 \\ \theta_n & \text{else} \end{cases} \tag{5}$$

At this point we have a new symbol sequence optimized for lower PAR. This new scheme needs two extra oversampled-IDFT operations at the transmitter. However, no modifications are needed at the receiver. IEEE 802.11a PHY layer parameters are depicted in Table 1.

Table 1. IEEE 802.11a PHY parameters

Parameter	Value
Information data rate	6,9,12,24,36,48 and 54 Mb/s
Modulation	BPSK –OFDM QPSK – OFDM 16-QAM – OFDM 64-QAM-OFDM
FEC code	Convolutional rate ½, constraint length 7
Code rate	1/2, 2/3, 3/4
Total number of subcarriers	52
Number of pilot subcarriers	4
OFDM symbol duration	4 μs
Guard interval	8 μs
Signal bandwidth	16.6 MHz

4. CLIPPING HIGH PEAKS

A soft limiter (SL) described below is used to clip the signal peaks. The nonlinear characteristics of an SL can be written as

$$y_k = \begin{cases} x_k & |x_k| \le A, \\ Ae^{j\phi_k} & |x_k| > A, \end{cases} \qquad k = 0,1,\ldots,4N-1 \tag{6}$$

where A is the clipping level and ϕ_k is the phase angle of the sample x_k and y_k is the clipped output sequence. The clipping ratio (CL) is defined as

$$\mathrm{CL} = 10\log_{10}\left(\frac{A^2}{E[|x_k|^2]}\right) \text{ dB.} \tag{7}$$

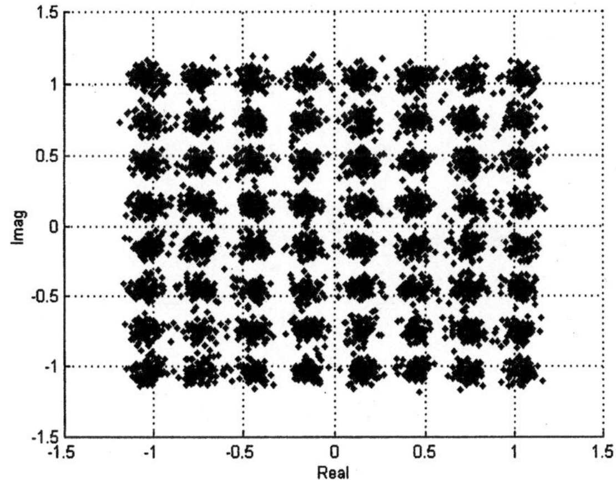

Figure 3. Signal distortion at 4 dB clipping of 48 Mbps signals

5. RESULTS

Simulation results were obtained for different transmit signals. As IEEE 802.11a transmit signals utilize four constellations (binary phase shift keying (BPSK), QPSK, 16-QAM and 64-QAM), simulations are performed for data rates involving these constellations except for BPSK. Clipping has little effect on BPSK OFDM. Figure 3 depicts the constellation of a 48 Mb/s transmitted signal passing through a SL with a 4dB back off. The distortion of the signal is clearly evident. Figure 4 shows the amplitude and phase corrected signal after 4dB clipping with correction factor $\alpha = 0.02$. The distortion of the signal reduces significantly depending on the value of the correction factor.

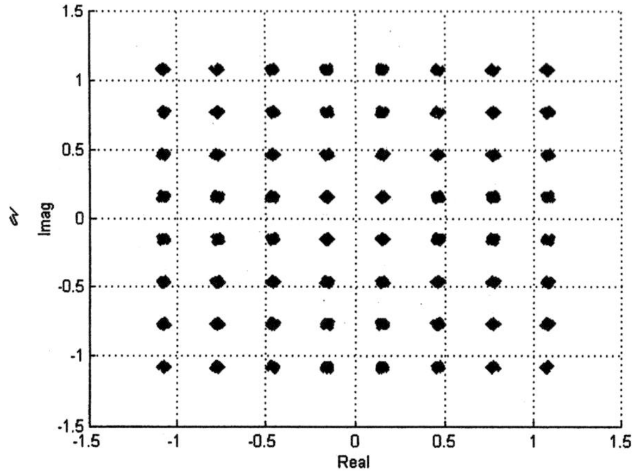

Figure 4. Corrected signal constellation at 4 dB clipping (48 Mbps and $\alpha = 0.02$)

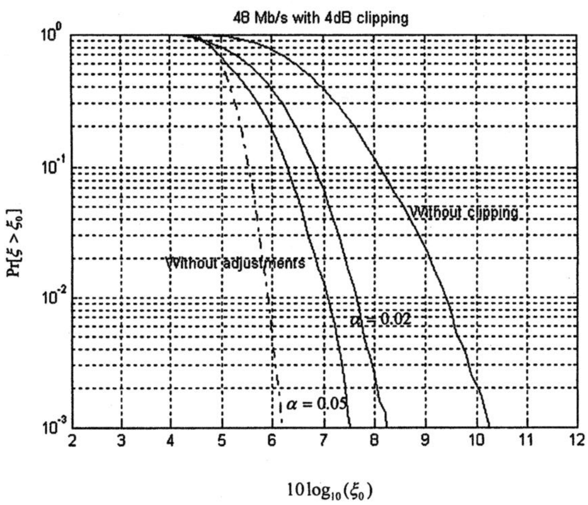

Figure 5. CCDF of the 48 Mbps signals with 4 dB clipping

Figure 5 depicts the complementary cumulative distribution function (CCDF) of the PAR of the 48 Mb/s signal. PAR is reduced by more than 4dB with the 4dB clipping, but with a severely distorted signal. Phase and amplitude correction reduces the distortion of the signal points but degrades the PAR statistics. A 3dB gain in PAR statistics is obtained when $\alpha = 0.05$, while this gain is around 2dB when $\alpha = 0.02$ at 10^{-3} of CCDF. Clipping with suitable phase and amplitude correction can achieve a significant PAR reduction.

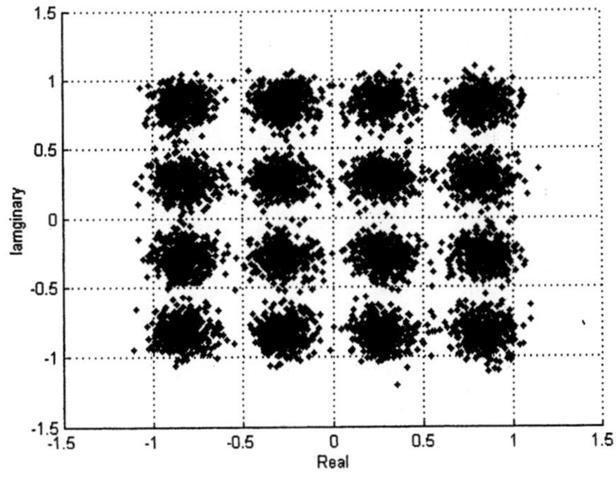

Figure 6. Signal constellation at 2 dB clipping of 36 Mbps signals

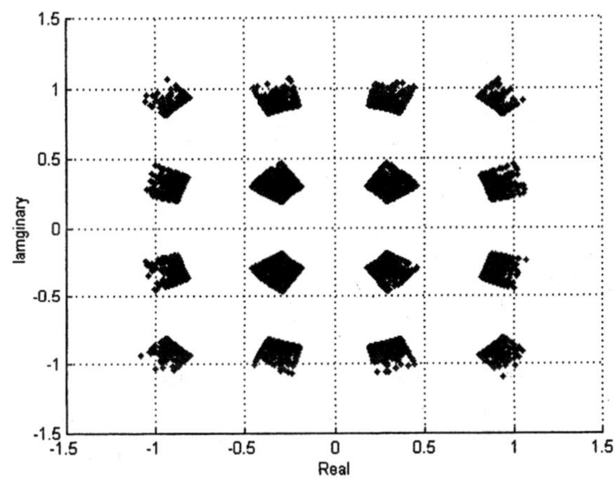

Figure 7. Corrected signal constellation at 2 dB clipping (36 Mbps and $\alpha = 0.1$)

We also examined the performance of other schemes. Figure 6 depicts the 36 Mb/s signal constellation after passing through a SL with 2dB clipping. As 16-QAM is less condensed than the constellation for 48 Mb/s or 54 Mb/s (64-QAM), we can allow for higher clipping ratios. Figure 7 shows the amplitude and phase corrected signal constellation. The amplitude and phase distortion is greatly reduced when the correction factor $\alpha = 0.1$. Figure 8 depicts the CCDF of PAR of the transmitted signals. The PAR statistics improve as before. The correction factor is chosen as $\alpha = 0.15$ and $\alpha = 0.1$ with clipping at 2dB. These parameters correspond to PAR statistics improvements of 3dB and 4dB respectively.

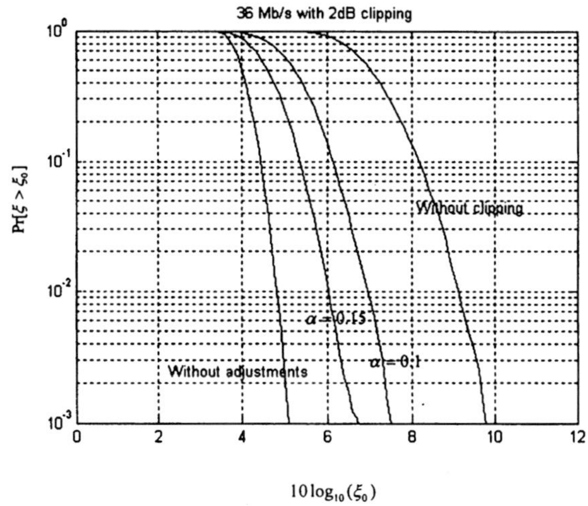

Figure 8. CCDF of the 36 Mbps signals with 2 dB clipping

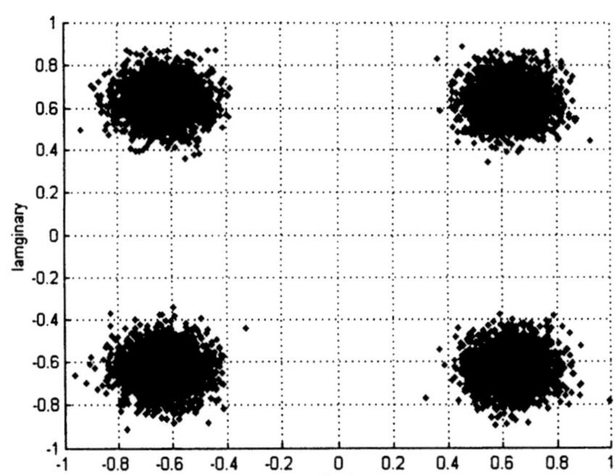

Figure 9. Signal distortion at 2 dB clipping of 12 Mbps signals

When the simulations are performed for 12 Mb/s signals we could observe far better performance improvements. Clipped 12 Mbps signal, amplitude and phase corrected signal and the CCDF are shown in Figure 9, Figure 10 and Figure 11 respectively. The constellation in 12 Mb/s signal is less dense than the two other schemes before and allows greater flexibility in selecting correction factor α. High values of α give out high PAR reduction. When the $\alpha = 0.2$ about 5dB gain is observed while this was more than 3dB when $\alpha = 0.1$

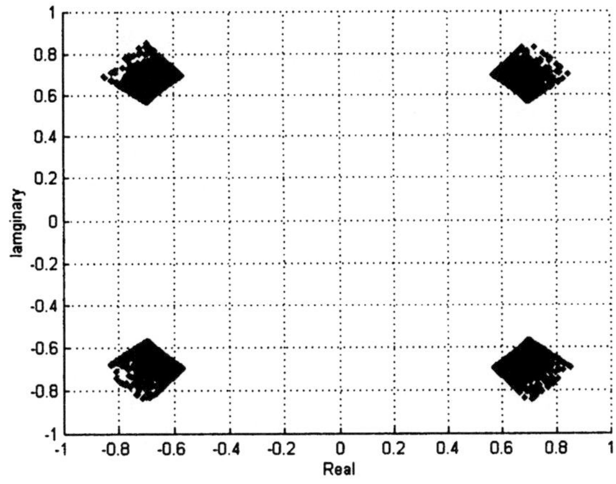

Figure 10. Corrected signal at 2 dB clipping (12 Mbps and)

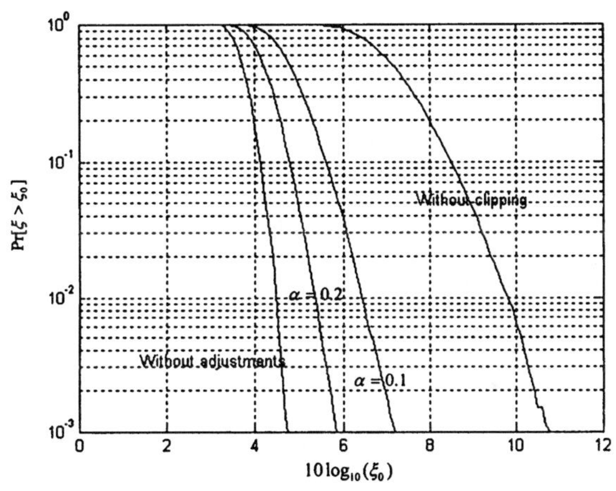

Figure 11. CCDF of the 12 Mbps signals with 2 dB clipping

Next we observe the power spectral density (PSD) of the transmitted signal when passing through a non-linear power amplifier. The input output relationship of power amplifiers can be modeled as

$$F[\rho] = \frac{\rho}{\left[1+\left(\dfrac{\rho}{A}\right)^{2P}\right]^{\frac{1}{2P}}} \tag{8}$$

$$\Phi[\rho] = 0$$

where A is the clipping level, ρ is the amplitude of the input signal and F and Φ denote the amplitude and the phase at the output respectively. The parameter P controls the smoothness of the transition from the linear region to the limiting or saturation region. When, $P \rightarrow \infty$, this model approximates the SL characteristics. P=3 approximates the solid state power amplifier. For PSD results, it is convenient to define the normalized bandwidth $B_n = f/(N\Delta f) = fT/N$, where T is the OFDM symbol duration.

In Figure 12, 12 Mb/s signal with 2dB clipping and correction factor $\alpha = 0.2$ is compared with an ordinary transmitted signal. The back-off of the power amplifier is set at different levels (3dB, 5dB, 7dB and 9dB). When the back off is very low (3dB) very slight improvement in OBN is observed. OBN is reduced by about 8dB when the back-off is at 5dB. Similarly, OBN reduces significantly with the increase in back-off at a faster rate compared to the ordinary signal. Therefore the back-off of the amplifier can be reduced significantly by using this technique.

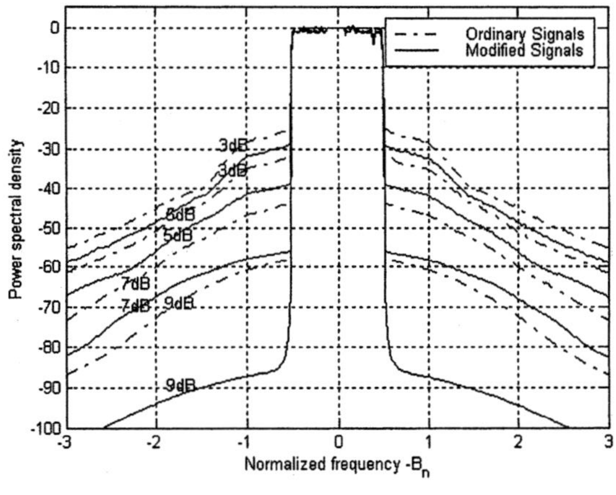

Figure 12. Power spectral density when passing through a non-linear amplifier

5.1 A measure of signal distortion

Effect of the clipping with amplitude and phase changes scheme on the signal distortion can be observed using the BER performance in a transmission channel. But in this section we introduce a new measure to evaluate the amount of distortion caused by the PAR reduction scheme. Let $\hat{X}_n$ be the received signal points. There are N such points in a N sub carrier OFDM system. A measure of distortion can be obtained by comparing the received signal ($\hat{X}_n$) with the transmitted signal X_n. Let Γ_c be the distortion after clipping,

$$\Gamma_c = \frac{1}{N}\sum_{n=0}^{N-1}\left|X_n - \hat{X}_n\right|^2, \quad n = 0,1,\ldots,N-1. \tag{9}$$

Then the gain after amplitude and phase adjustment ς_{apa} can be expressed as,

$$\varsigma_{apa} = 10\log_{10}\left[\frac{\Gamma_{apa}}{\Gamma_c}\right] \text{dB} \tag{10}$$

where Γ_{apa} is the distortion of the signal after amplitude and phase adjustments.

Table 2. Distortion of the clipped signal

Data rate	Clip level	α	Γ_c	Γ_{apa}	ς_{apa}	PAR reduction
48 Mbps	4 dB	0.02	0.0290	0.0119	-2.28 dB	2 dB
		0.05	0.0292	0.0212	-5.63 dB	3 dB
36 Mbps	2 dB	0.10	0.1082	0.0684	-4.34 dB	3 dB
		0.15	0.1014	0.0837	-7.55 dB	4 dB
12 Mbps	2 dB	0.10	0.0926	0.0626	-4.89 dB	3 dB
		.020	0.0906	0.0877	-14.94 dB	5 dB

Table 2 shows the distortion and the gain after amplitude the phase correction of a clipped OFDM signal. It also shows the achievable PAR reduction. Selection of initial clipping ratio and the correction factor depends on the signal constellation being used. Therefore, by selecting a suitable clipping level and a proper phase and amplitude correction factor significant PAR reduction is achieved without causing significant BER degradation. This is a desired feature in portable devices in WLANs, where power efficient transmitter power amplifiers are essential.

6. CONCLUSION

A technique based on clipping with amplitude and phase changes to reduce the PAR of OFDM based WLAN signals defined in the IEEE 802.11a physical layer is presented in this chapter. The proposed technique is capable of reducing the PAR by 3-4dB by selecting a suitable clipping level and amplitude and phase correction factor. It can be implemented with a slight increase in the complexity at the transmitter. This involves with insertion of two additional DFT operations and soft limiter at the transmitter. The receiver remains standard compliant.

REFERENCES

[1] IEEE, " Std 802.11a-1999 (supplement to IEEE Std 802.11-1999), Part II: Wireless LAN medium access control (MAC) and physical layer (PHY) specifications: High speed physical layer in 5 GHz band," IEEE, tech. rep. Sep. 1999.

[2] D. W. Bennett, P. B. Kenington, and R. J. Wilkinson, "Distortion Effects of Multicarrier Envelope Limiting," *IEE Proc. Commun.*, vol. 144, pp. 349-356, 1997.

[3] X. Li and L. J. Cimini, "Effects of clipping and filtering on the performance of OFDM," *IEEE Commun. Letts.*, vol. 2, pp. 131-133, 1998.

[4] H. Ochiai and H. Imai, "On the clipping for peak power reduction of OFDM signals," *IEEE GLOBECOM*, 2000.

[5] J. Armstrong, "New OFDM peak-to-average power reduction scheme," presented at IEEE VTC, 2001.

[6] D. Wulich and L. Goldfeld, "Reduction of peak factor in orthogonal multicarrier modulation by amplitude limitting and coding," *IEEE Trans. Commun.*, vol. 47, pp. 18-21, 1999.

[7] D. Kim and G. L. Stuber, "Clipping Noise Mitigation for OFDM by Decision-Aided Reconstruction," *IEEE Commun. Letts.*, vol. 3, pp. 4-6, 1999.

[8] R. O'Neili and L. B. lopes, "Performance of amplitude limited multitone signals," *Intl. Conf. Commun.*, 1994.

[9] D. J. G. Mestdagh, P. Spruyt, and B. U. Brain, "Analysis of clipping effect in DMT-based ADSL systems," *IEEE Intl. Conf. Commun.*, 1994.

[10] R. Gross and d. Veeneman, "SNR and spectral properties for a clipped DMT ADSL signal," *IEEE Intl. Conf. Commun.*, 1994.

[11] M. Friese, "On the degradation of OFDM-signals due to peak-clipping in optimally predistorted power amplifiers," *IEEE GLOBECOM*, Sydney, Australia, 1998.

[12] E. Bogenfeld, R. Valentin, K. Metzger, and W. Sauer-Greff, "Influence of Nonlinear HPA on Trellsi-Coded OFDM for Terrestrial Broadcasting of Digital HDTV," IEEE *GLOBECOM '93*, 1993.

[13] G. Santella and F. Mazzenga, "A hybrid analytical-simulation procedure for performance evaluation in M-QAM-OFDM schemes in presence of nonlinear distortions," *IEEE Trans. Veh. Technol.*, vol. 47, pp. 142-151, 1998.

[14] P. Shelswell and M. C. D. Maddocks, "Digital signal transmission system using frequency division multiplex," in *U. S. Patent*, No. 5, 610, 908, 1997.

[15] C. Tellambura, "Phase optimization criterion for reducing peak-to-average power ratio in OFDM," *Elect. Letts.*, vol. 34, pp. 169-170, 1998.

Chapter 8

A PROPOSED HANGUP FREE AND SELF-NOISE REDUCTION METHOD FOR DIGITAL SYMBOL SYNCHRONIZER IN MFSK SYSTEMS

C..D.Lee and M.Darnell
Institute of Integrated Information Systems, School of Electronic and Electrical Engineering, University of Leeds LS2 9JT

Abstract: The contribution of this paper concerns digital symbol synchronizers for MFSK systems. Methods are proposed to achieve hangup free operation and to reduce self-noise. Symbol synchronization hangup can be detected and acquisition time can be reduced if all the samples per symbol from the matched filter output are used. A phase-modulated-M-ary-frequency-shift-keying (PM-MFSK) scheme is introduced to reduce self-noise due to random data. In the analysis, the timing error variances of the symbol synchronizers developed are derived. The synchronizers are evaluated via acquisition time required and normalized timing error variance under additive white Gaussian noise (AWGN) channel conditions.

Key words: Symbol synchronization, MFSK, Synchronization hangup

1. INTRODUCTION

Most symbol synchronizers have been analyzed for M-ary phase-shift-keying (MPSK) or quadrature amplitude modulation (QAM) [1]-[4]. The implementation of a symbol synchronizer for MFSK is slightly different from MPSK because the former has M basis functions, which means each signaling tone is a basis function. Hence, we need to implement M matched filters in parallel so that each can be matched to a particular signaling tone. Coherent detection is used here and perfect carrier synchronization is assumed. Symbol synchronization hangup and self-noise are the common problems faced by symbol synchronizers. Synchronization hangup is a phenomenon that occurs when normalized timing error is close to 0.5 (i.e.

the unstable equilibrium point); it will tend to move to a stable point but, in doing so, its evolution will be governed by noise since the steering force S-curve is small. As a result, the occurrence of a hangup gives rise to a long acquisition time. Both thermal noise and self-noise contribute to timing error variance. The former dominates at small SNR and the latter dominates at high SNR. The presence of self-noise is due to the random data. This paper is divided into sections as follows: section II describes the operation of the digital symbol synchronizer; section III shows how synchronization hangup can be detected and corrected; section IV illustrates how self-noise can be reduced; the performance of the proposed synchronizer is analyzed in section V and evaluated in section VI; finally, in section VII, we conclude the paper.

2. DIGITAL SYMBOL SYNCHRONISER

After the symbol synchronizer has obtained the initial estimate from a known sequence, the detection process can be started. Consequently, the estimated sequence $\hat{\boldsymbol{c}}$ can be used to remove data dependency. This method is called decision-directed (DD) synchronization algorithm and is given by

$$\hat{\tau} = \arg\max_{\tau}\, p\left(r \middle| \tilde{\mathbf{c}} = \hat{\mathbf{c}}, \tau\right) \qquad (1)$$

The log-likelihood function is given by [2]

$$L\left(r \mid \tilde{\tau}, \tilde{\mathbf{c}}\right) = \sum_{m=0}^{(L_0-1)N-1} \mathrm{Re}\left\{r(mT_s)s^*(mT_s,\tilde{\tau})\right\} - \frac{1}{2}\left|s(mT_s,\tilde{\tau})\right|^2 \qquad (2)$$

where L_0 is the observation length in symbols. The $s(mT_{s,}\tilde{\tau})$ is the trial signal given by

$$s(mT_s,\tilde{\tau}) = \sum_{k=-\infty}^{\infty} \hat{c}_k(mT_s)\cdot g\left(mT_s - kT - \tilde{\tau}\right) \qquad (3)$$

where $\hat{\mathbf{c}}$ represents estimated signaling data sequence, $g(t)$ is a real-valued pulse, and $\tilde{\tau}$ is the generic channel delay. However, the second term is usually dropped because the integral of the term has only a small variation with $\tilde{\tau}$ [2]-[3]. Thus, the approximate log-likelihood function is given by

$$L(r \mid \tilde{\tau}, \tilde{\mathbf{c}}) \approx \sum_{m=0}^{(L_0-1)N-1} \operatorname{Re}\left\{ r(mT_s) s^*(mT_s, \tilde{\tau}) \right\} \tag{4}$$

Substituting in (4) from (3), (4) can be written as

$$L(r \mid \tilde{\tau}, \tilde{\mathbf{c}}) \approx \sum_{m=0}^{(L_0-1)N-1} \sum_{k=-\infty}^{\infty} \operatorname{Re}\left\{ r(mT_s) \cdot \hat{c}_k^*(mT_s) \cdot g(mT_s - kT - \tilde{\tau}) \right\} \tag{5}$$

Here we make an approximation that is valid, especially when the observation interval is much longer than the duration of the pulse $g(t)$ [1]. It consists of increasing the summation over m to $\pm\infty$ while restricting the summation over k from 0 to L_0 -1. Hence, (5) can be approximated to

$$\sum_{m=0}^{(L_0-1)N-1} \operatorname{Re}\left\{ r(mT_s) \cdot s^*(mT_s, \tilde{\tau}) \right\} \approx \sum_{k=0}^{L_0-1} y(kT + \tilde{\tau}, \hat{c}_k) \tag{6}$$

where $y(t)$ is the response of $r(t)$ to the matched filter $s(t)$ and is given by

$$y(kT + \tilde{\tau}) = \sum_{m=-\infty}^{\infty} \operatorname{Re}\left\{ r_k(mT_s) \cdot \hat{c}_k^*(mT_s) \cdot g(mT_s - kT - \tilde{\tau}) \right\} \tag{7}$$

800

Thus,

$$L(r \mid \tilde{\tau}, \tilde{\mathbf{c}}) \approx \sum_{k=0}^{L_0-1} y(kT + \tilde{\tau}, \hat{c}_k) \tag{8}$$

A DD ML timing parameter estimator is a timing parameter estimator that maximises (8). The log-likelihood function for DD-MDS is defined as

$$\sum_{k=0}^{L_0-1} y(kT + \tilde{\tau}, \hat{c}_k) \approx \sum_{k=0}^{L_0-1} \max_{{}^i\tilde{c}_k} \left(y(kT + \tilde{\tau}, {}^i\tilde{c}_k) \right) \quad \text{for } i = 1, \ldots, M \tag{9}$$

where ${}^i\tilde{c}_k$ is the ith signaling tone in the alphabet set at k^{th} interval. The DD ML symbol synchronizer here is implemented differently from the one for MPSK [1]-[4] because, in MFSK, the generic data waveform $\tilde{c}$ is not constant within a symbol period. The basic concept of the tracking-loop implementation is to search for the τ that causes the differential of the approximate log-likelihood function to be zero:

$$\frac{\partial \Lambda(\tilde{\tau})}{\partial \tilde{\tau}} = 0 \tag{10}$$

The structure of the modified tracking-loop MDS (MTMDS) is shown in Figure 1. In tracking-loop MDS (TMDS), the synchronization epoch is determined by the position of the zero in the differential buffer. Consequently, following from (9) and (10) the ML estimate for the MTMDS with exponential memory is the $\tilde{\tau}$ that satisfies (11)

$$\frac{\partial}{\partial \tilde{\tau}}\left[\frac{1}{L_a}\sum_{k=0}^{\infty}\left(\frac{L_a-1}{L_a}\right)^k \max_{{}^i\tilde{c}_k}\left(y(kT+\tilde{\tau},{}^i\tilde{c}_k)\right)\right]=0 \tag{11}$$

$$\text{for } i=1,\dots,M$$

where L_a is the averaging filter memory. The averaging filter is an infinite impulse response (IIR) filter with feedback gain of $(L_a\text{-}1)/L_a$. The implementation of the averaging filter requires the assumption that the sampling rate is very close to an integer multiple of symbol rate. The interpolator is used for fine timing adjustment. According to (11), the decision-directed unit (DDU) chooses the matched filter output that has the maximum amplitude. By choosing the maximum amplitude, the DDU has indirectly made a symbol decision at every sampling instant; those symbol decisions are stored in the symbol buffer. The timing error detector is a differential operator. The loop filter updates the timing estimate as follows:

$$\tau(k+1)=\tau(k)+\gamma e(k) \tag{12}$$

where $e(k)$ is the timing error signal at the kth instant and γ is related to the noise equivalent bandwidth B_L as follows [1]

$$B_L T = \frac{\gamma A}{2(2-\gamma A)} \tag{13}$$

where A is the slope of S-curve when the timing error is equal to zero. The estimated timing epoch is produced at the symbol rate. The integer part of the estimated timing epoch is used to select one of the symbol decisions from symbol buffer as the estimated symbols are obtained at symbol rate; the fractional part is used for fine timing adjustment. The difference between the novel synchroniser and the conventional one is that the latter decimates the matched filter output to 1 or 2 samples per symbol and it does not have a DDU.

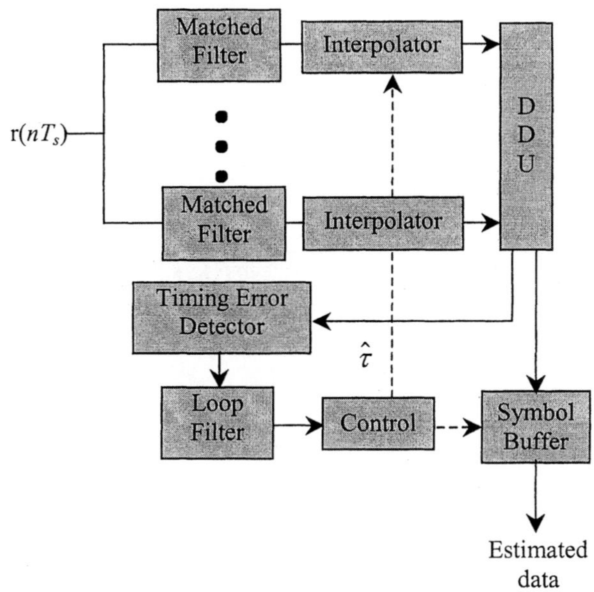

Figure 1. The structure of MTMDS

3. HANGUP DETECTION AND CORRECTION

Figure 2 shows the S-curve for different numbers of samples per symbol. From Figure 2, we notice that different numbers of samples per symbol will have different hangup regions. A vertical line drawn from the critical points in Figure 2 identifies the hangup region and non-hangup region. The critical point is empirically found to be at

$$\left|\frac{\delta}{T}\right| = \frac{1}{2N}\left(N-K\right) \tag{14}$$

where δ is the timing error, T is the symbol period, N is the number of samples per symbol and K is the distance (in terms of number of samples) on the x-axis (i.e. time axis) between the 2 points where the differential is calculated. Thus, we know the hangup region is at

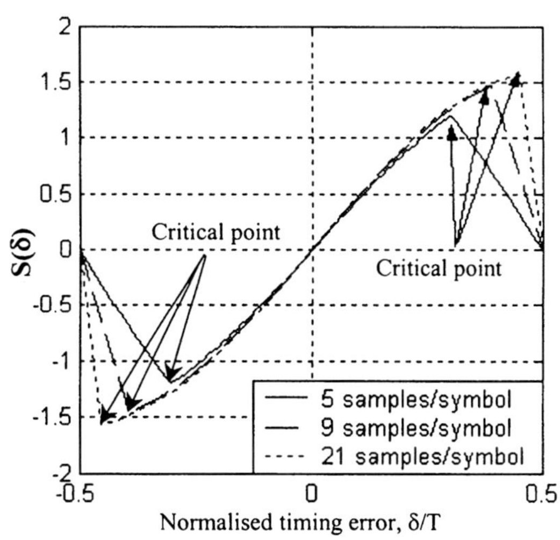

Figure 2. S-curve for different numbers of samples/symbol.

$$\left|\frac{\delta}{T}\right| > \frac{1}{2N}\left(N-K\right) \tag{15}$$

In order to detect a hangup region, we need to find the difference between the current sampling point and the position of the peak. This is illustrated in Figure 3. Note that this could be done only if all the N samples were preserved after the matched filter. This requires a combination of peak search MDS [5]-[6] and MTMDS, now we called hybrid MDS (HMDS). The structure of HMDS is shown in Figure 4. The peak search unit simply searches for the position of the peak, and then the timing error detector will calculate the distance between its current sampling point and the peak position. If (15) is satisfied, then a hangup is declared and the current sampling point is immediately changed to the position of the peak. If hangup is not declared, then the operation is similar to MTMDS.

4. SELF-NOISE REDUCTION

Self-noise is caused by the inter-symbol interference (ISI) between the raised-cosine pulses. The self-noise in the symbol synchronizer increases when the rolloff of the raised cosine pulse decreases. For a rolloff of 1.0, the main contribution of self-noise is due to the previous symbol and the next symbol. Thus, by assuming that the pulse duration greater than $2T$ is negligible, there are 3 possible cases:

i) If all three symbols have the same polarity, then no timing information is available.
ii) If only one of the symbol (i.e. either the previous or the next symbol) has similar polarity to the current symbol then wrong timing information is obtained.
iii) If both the previous and the next symbols have the opposite polarity to the current symbol, then the timing information is accurate, provided that the timing error is small.

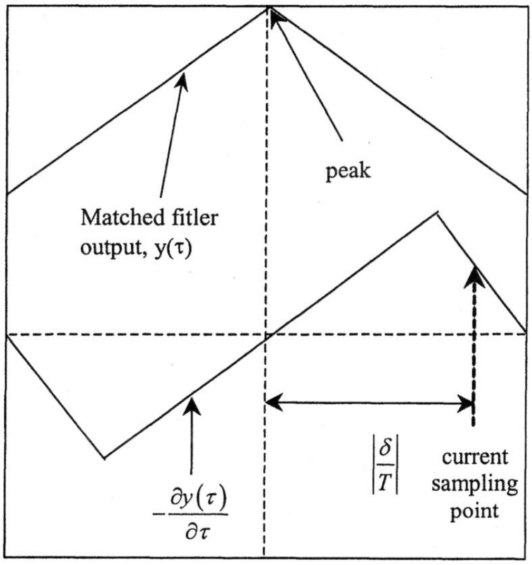

Figure 3. The relationship between the matched filter output and S-curve.

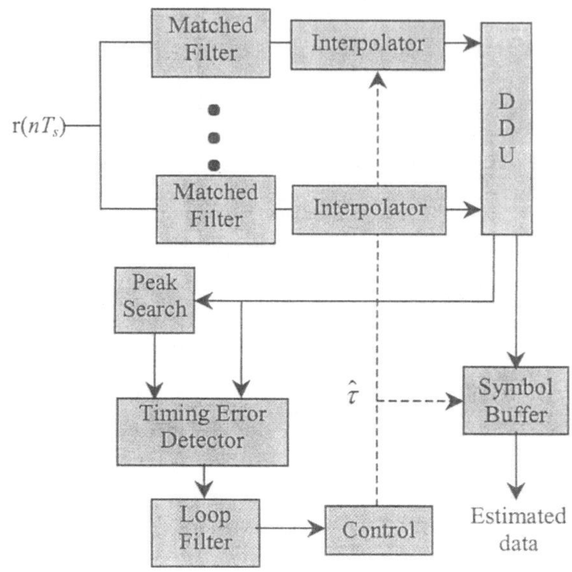

Figure 4. The structure of HMDS.

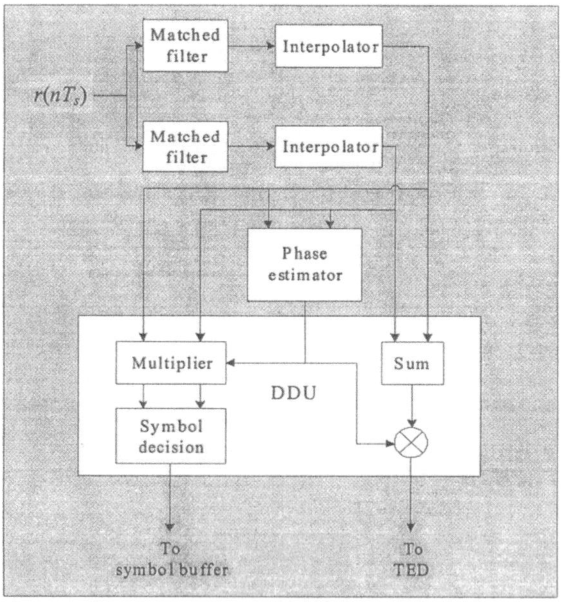

Figure 5. The structure of SCMDS.

The option (iii) is what we strive to achieve. In MFSK, the data is conveyed via selecting the frequency tone. For each MFSK symbol, we invert the phase of the frequency tone so that the phase of successive MFSK symbols alternates between 0 and π. We term this phase-modulated-M-ary-frequency-shift-keying (PM-MFSK). Since the phase modulation is independent of the data, option (iii) is satisfied for all possible data sequences. Since this synchronizer reduces self-noise, we term this self-noise cancellation MDS (SCMDS). Minor modification has to be made to the symbol synchronizer when PM-MFSK is applied. The modification made is shown in Figure 5. The phase-modulated output of the matched filter can be removed by using a phase estimator. Non-linear device, such as a squarer or an absolute function, can also be used but it introduces more thermal noise into the system. In order to reduce normalized timing error variance in the steady state, the contribution of ISI from the previous symbols must be the same as the next symbols so that the current pulse shape remains symmetrical and the differential at the optimum point of a symmetrical pulse is zero (i.e. the error signal is zero). Therefore, all the matched filter outputs have to be summed since the previous pulses and the future pulses might be located in different matched filter branches. This means that we now have a non-decision-directed (NDD) ML symbol synchronizer. Although this introduces more thermal noise into the system, it is unavoidable. Consequently, self-noise cancellation MTMDS (SCMTMDS) and self-noise cancellation HMDS (SCHMDS) were developed by applying SCMDS to MTMDS and HMDS, respectively.

5. PERFORMANCE ANALYSIS

Normalized timing error variance is given as [1]

$$\sigma_t^2 = \frac{1}{T^2} \sum_{m=-\infty}^{\infty} R_n(m)\eta(m) \tag{16}$$

where $R_n(m) = E\{n(k+m)n(k)\}$ is the autocorrelation function of the noise and $\eta(m)$ is the convolution of $h(k)$ with $h(-k)$, where $\boldsymbol{h}$ is the loop filter response to $n(k)$, i.e.

$$\eta(m) = \sum_{i=-\infty}^{\infty} h(i)\, h(i-m) \tag{17}$$

After some manipulation, it is found that [1]

$$\eta(m) = \frac{\gamma}{A(2-\gamma A)} \ (1-\gamma A)^{|m|} \tag{18}$$

where A is the slope of the S-curve at a timing error equal to zero.

$$A = -q''(0) \tag{19}$$

and for SCMDS will be

$$A_S = -2q''(0) \tag{20}$$

and $q(t) = g(t) \otimes g_{MF}(-t)$ is the convolution of the root-raised cosine pulse with the matched filter response. Then, substituting into (16) yields

$$\sigma_t^2 = \frac{\gamma}{A(2-\gamma A)T^2} \sum_{m=-\infty}^{\infty} R_n(m) \ (1-\gamma A)^{|m|} \tag{21}$$

Substituting (13) into (21), we have

$$\sigma_t^2 = \frac{2B_L T}{A^2 T^2} \sum_{m=-\infty}^{\infty} R_n(m) \ (1-\gamma A)^{|m|} \tag{22}$$

The computation of the normalized timing error variance σ_t^2 requires a knowledge of the autocorrelation $R_N(m)$ of the noise. Since the self-noise $n_s(k)$ and the thermal noise $n_t(k)$ are uncorrelated, we have

$$R_n(m) = R_t(m) + R_s(m) \tag{23}$$

Now considering on $R_t(m)$: the noise from different matched filter branches is assumed to be uncorrelated; thus

$$R_n(m) = \begin{cases} E\{n_t^2(t)\} & m = 0 \\ 0 & m \neq 0 \end{cases} \tag{24}$$

To evaluate the noise variance, we need the Fourier transform of $g'(-t)$ which is found to be $-j2\pi f\, G^*(f)$, where $G(f)$ is the Fourier transform of $g(t)$. Thus, the noise variance is given by [1]

$$\sigma_n^2 = E\left\{n_t^2(t)\right\} = \frac{N_0}{2} \cdot 4\pi^2 \int_{-\infty}^{\infty} f^2 \,|\, G(f) \,|^2 \; df \tag{25}$$

$N_0/2$ is the double-sided power spectral density of additive white Gaussian noise. Note that the noise variance for SCMDS is M time larger than (25) since the noise from the M matched filters branches is summed.

$$\tilde{\sigma}_n^2 = M \cdot E\left\{n_t^2(t)\right\} = \frac{MN_0}{2} \cdot 4\pi^2 \int_{-\infty}^{\infty} f^2 \,|\, G(f) \,|^2 \; df \tag{26}$$

Assuming that the self-noise is negligible, and collecting terms from (22)-(25), the normalized timing error variance is given by

$$\sigma_t^2 = \frac{N_0 B_L}{A^2 T} \cdot 4\pi^2 \int_{-\infty}^{\infty} f^2 \left|G(f)\right|^2 \; df \tag{27}$$

Then, substituting A from (19), we have

$$\sigma_t^2 = \frac{N_0 B_L}{q''^2(0) T} \cdot 4\pi^2 \int_{-\infty}^{\infty} f^2 \left|G(f)\right|^2 \; df \tag{28}$$

and recognizing $q(t) = g(t) \otimes g_{MF}(-t)$, it can be seen that $q''(t)$ has the Fourier transform $-4\pi^2 f^2 \left|G(f)\right|^2$ and, hence, $q''(0)$ can be written as

$$q''(0) = -4\pi^2 \int_{-\infty}^{\infty} f^2 \left|G(f)\right|^2 df \tag{29}$$

Also the signal energy is given by [1]

$$E_s = \int_{-\infty}^{\infty} \left|G(f)\right|^2 df \tag{30}$$

substituting (29) and (30) into (28) produces

$$\sigma_t^2 = \frac{B_L T}{4\pi^2 \xi} \cdot \frac{1}{E_s / N_0} \tag{31}$$

where ξ is an adimensional coefficient given by [1]

$$\xi = T^2 \frac{\int_{-\infty}^{\infty} f^2 |G(f)|^2 \, df}{\int_{-\infty}^{\infty} |G(f)|^2 \, df} \tag{32}$$

It is also given that [1]

$$B_L T = \frac{1}{2L_0} \tag{33}$$

Hence, by substituting (33) into (31), we have

$$\sigma_t^2 = \frac{1}{8\pi^2 \xi L_0} \cdot \frac{1}{E_s / N_0} \tag{34}$$

which is the well-known modified Cramer-Rao bound (MCRB) [1]. Hence, MTMDS converges to the MCRB when self-noise is negligible. Using the same procedure, but replacing (19) with (20) and (25) with (26), we have the normalized timing error variance for SCMDS

$$\sigma_t^2 = \frac{M}{4} \left(\frac{1}{8\pi^2 \xi L_0} \cdot \frac{1}{E_s / N_0} \right) \tag{35}$$

which differs from MCRB by a factor of *M/4*.

6. PERFORMANCE EVALUATION

The performances of the 4 developed symbol synchronizers MTMDS, HMDS, SCMTMDS and SCHMDS were evaluated and compared using performance measures such as acquisition time required and normalized timing error variance. Only 5 samples per symbol were used for an alphabet size of 2 and the channel was AWGN. L_0 was chosen to be 100.

6.1 Acquisition time required

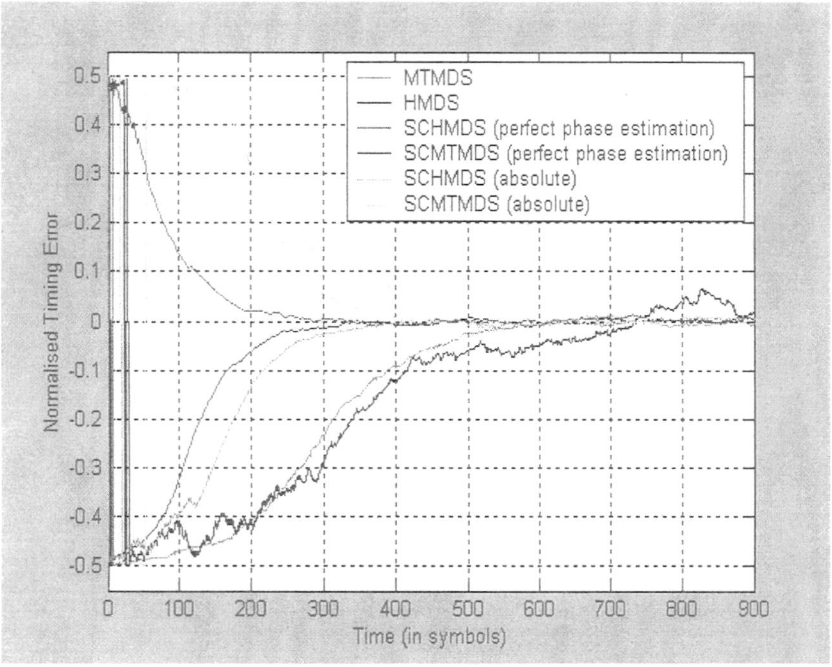

Figure 6. Acquisition time required.

In the simulation, we let the sampling start at the worst possible sampling position, which is when the normalized timing error is equal to ±0.5. In other words, sampling at the boundary of the symbols. Random data was used and E_b/N_0 = 15 dB. The result is shown in Figure 6. The convergence rate of MTMDS and HMDS are poor. HMDS requires not only a longer acquisition time than MTMDS, but it also shows a larger timing error fluctuation than MTMDS at the tracking stage; this is due to the wrong timing information being provided by the repeated symbols. Therefore, HMDS performs worse than MTMDS because it is severely affected by self-noise. The pulses for SCMDS can be rectified using a phase estimator or an absolute function. In both perfect phase estimation case and absolute function case, SCMTMDS and SCHMDS have faster acquisition time than MTMDS and HMDS. SCHMDS has faster convergence rate than SCMTMDS since the hangup region is avoided in SCHMDS. The absolute function degrades the performance of SCMTMDS and SCHMDS because it introduces more noise into the system. The results show that the proposed anti-hangup algorithm only works correctly when the self-noise is removed. The reason for choosing $E_b/N_0 > 10$ dB is because that is the range of practical E_b/N_0. Although the $E_b/N_0 = 15$ dB is chosen, other E_b/N_0 that were > 10 dB show similar trends.

6.2 Normalized timing error variance

When the loop bandwidth of the synchronizer is much smaller than the symbol rate, the statistical properties of the timing error detector output signal under open-loop conditions remain valid when the loop is closed [1]. The timing error variance shown in Figure 7 is normalized by the symbol period:

$$\sigma_t^2 = \frac{(\tau - \hat{\tau})^2}{T^2} \tag{36}$$

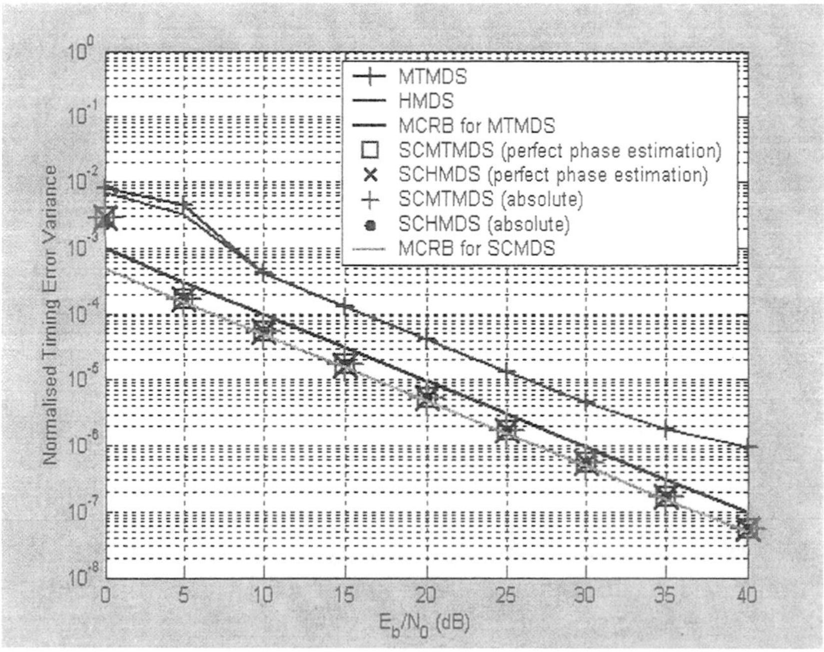

Figure 7. Normalized timing error variance.

Note that the MCRB for MTMDS is different from the MCRB for SCMDS. For M=2, the MCRB for SCMTMDS is 3 dB lower than the MCRB for MTMDS. HMDS and MTMDS are severely affected by self-noise due to repeated symbols since they are approximately 6 dB away from the MCRB for MTMDS. MTMDS is affected by self-noise. Both SCHMDS and SCMTMDS converge to the MCRB for SCMDS and coincide with the bound. This shows that they are almost self-noise free and the timing error is only due to thermal noise. The performance of SCMDS using an absolute function is the same as the performance of SCMDS with perfect phase estimation at high E_b/N_0. The performance of SCMTMDS and SCHMDS are better than MTMDS and HMDS because they have very low self-noise. Consequently, the SCMTMDS and SCHMDS not only have fast acquisition property but also low normalized timing error variance. Both SCHMDS and SCMTMDS have the lowest normalized timing error variance but SCHMDS has the fastest acquisition time. Therefore, it is concluded here that the best symbol synchronizer in AWGN channel conditions among the compared synchronizers is SCHMDS.

7. CONCLUDING REMARKS

An improved version of TMDS is developed. The advantages of MTMDS over TMDS include an ability to implement fine timing adjustment and lower complexity, since only 1 sample per symbol is needed to derive the timing estimates; it works even when T/T_s is irrational. Nevertheless, MTMDS is quite similar to the symbol synchronizer described in [1], except the latter was not developed for MFSK systems and it does not have a DDU. HMDS was developed to remove synchronization hangup, but the objective was not achieved because it is severely affected by self-noise. Nevertheless, SCHMDS successfully removes synchronization hangups when the self-noise is removed. Both SCMTMDS and SCHMDS are almost self-noise free but they are susceptible to thermal noise because the summation of all matched filter outputs increases the thermal noise variance by a factor M, where M is the alphabet size. it is concluded here that the best symbol synchronizer in AWGN channel conditions among the compared synchronizers is SCHMDS.

REFERENCES

[1] U. Mengali, A. N. D'Andrea, *Synchronization Techniques for Digital Receivers*, Plenum Press, 1997.

[2] H. Meyr, M. Moeneclaey, S. A. Fechtel, *Digital Communication Receivers: Synchronization, Channel Estimation, and Signal Processing*, John Wiley & Sons, Inc., 1998.

[3] M. Oerder, "Derivation of Gardner's timing-error detector from the maximum likelihood principle," *IEEE Trans. on Commun.,* Vol 35, No. 6, June 1987.

[4] F. M. Gardner, "A BPSK/QPSK Timing-Error Detector for Sampled Receivers," *IEEE Trans. Commun.*, vol. COM-34, no. 5, pp. 423-429, May 1986.

[5] S. M. Brunt, *Comparison of Modulation Derived Synchronisation and Conventional Symbol Synchronisation Techniques in Time Dispersive Channels*, PhD Thesis, University of Leeds, April 1998.

[6] C. D. Lee, S. Brunt, M. Darnell, "Comparison of Various Types of Modulation-Derived Synchronisers under Low Sampling Rate Conditions", *5th International Symposium on Communication Theory and Applications,* Ambleside, July 1999.

Chapter 9

A CHANNEL SOUNDER FOR LOW-FREQUENCY SUB-SURFACE RADIO PATHS

D. Gibson and M. Darnell

Institute of Integrated Information Systems, University of Leeds, LEEDS, LS2 9JT, UK. D.Gibson@caves.org.uk, M.Darnell@elec-eng.leeds.ac.uk

Abstract: In an application to measure subterranean radio propagation, a wide-band low-frequency channel sounder makes use of a modified pseudo-random binary sequence with an imperfect (non-impulsive) auto-correlation function (ACF). The transmitting antenna for this system is essentially an induction loop, and the extreme wideband nature of the system requires this to be operated untuned. The resultant low efficiency is countered by signal-averaging techniques at the receiver, requiring accurate synchronisation to the transmitter, which is achieved using a code- or delay-locked loop. Accurate system identification and synchronisation is possible because cross-correlation is carried out between the non-ideal transmitted sequence and its inverse reference. A simple algorithm is presented for calculating the inverse of an arbitrary sequence under the operation of cross-correlation, and some properties of inverse sequences are discussed.

Key words: system identification, sequence design, inverse sequences channel-sounding, sub-surface, LF, VLF.

1. INTRODUCTION: SUB-SURFACE RADIO

This paper describes an application of DSP techniques to sub-surface digital radio communications. In particular it describes a system for performing channel-sounding experiments in the VLF, voice and LF bands (300Hz to 300kHz). A discussion of sub-surface radio systems, with reference to the peculiarities of the propagation, antenna requirements and the development of an adaptive digital communications system has already been given [1]. To recap, the features that characterise these systems are

a) Communications through conductive media over distances of only a few skin depths – but not restricted to near field.
b) Small, portable antennas. (But large fixed installations are also used by the mining industry).
c) Restricted choice of frequency. (Interference, noise, other primary channel users).
d) Optimum frequency and orientation of antenna depends on many external factors, including transmission distance and ground conductivity, as well as noise and co-channel interference.

Induction radio has been used for communications in the mining industry, and also by cavers and pot-holers. In addition to uses for speech communications, radiolocation and cave surveying beacons make use of the known behaviour of the near-field lines, so accurate results are limited to under a skin depth, although deeper work is possible when a more accurate propagation model is used [2]. Archaeological and geophysical applications include the remote measurement of ground conductivity by analysing the phase of the induced fields from an induction loop.

Subterranean radio systems differ from the ELF (30–300Hz) systems, which have been developed for submarine communications, because the latter operate in the far field – at 300Hz the transition distance is at 160km in air, but only 15m in seawater. The analysis of such systems involves concepts that are not applicable to a subterranean radio system. Cave radios differ from mine communications systems because they usually use small, portable antennas and a low power, whereas a mining application can use a large, powerful, fixed antenna, and needs to address the problem of transmission through conductive over-burden to deep workings.

2. CHANNEL SOUNDING

2.1 Uses of channel sounding

Propagation in a conductive medium, although well studied, can still be difficult to describe succinctly. Channel sounding, as well as providing the practical means to determine the optimum operation of a communication system, will allow us to test the theory and to develop some simple and instructive models of the propagation. In addition, the same equipment will be used to make measurements of background noise and interference at a high resolution, in a similar manner to that reported by Laflin [3]. Because channel sounding is a swift, automated exercise we will be able to make measurements at many locations underground, under a range of different conditions. It is planned to use specially-designed datalogging equipment to capture data at remote locations over a period of many months.

2.2 Measurement Spectrum

Almost the entire LF band (30–300kHz) has been used for subsurface communications at one time or another, but most through-rock communications takes place at 80–100kHz in the UK and Europe. In areas with much drier rock, 185kHz is a popular choice of frequency. Beacon communications (e.g. for radiolocation or telemetry) tend to be in the voice frequency band at 0.3–3kHz. We might, eventually, wish to evaluate the spectrum over three decades, from 300Hz to 300kHz, although an initial investigation of propagation, noise and interference will be restricted to the band from 15kHz to 240kHz (a range of 16:1). This range encompasses several existing amateur and commercial designs of beacon and radio [4].

2.3 Experimental Method

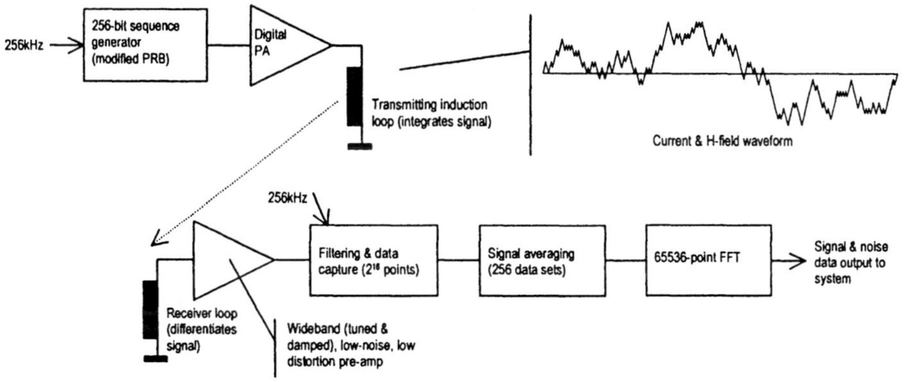

Figure 1. Block diagram of the prototype channel-sounding system.

The sounding signal is a wideband binary sequence. The original intention was to use a high-speed data-capture board and a laptop PC to test the system in the field. However, problems arose because of the power consumption and the large amount of real-time data – 32Mbytes per sounding – that needs to be captured. Instead, some dedicated hardware is being developed around an industrial datalogger. The data is reduced in real time to 512kbytes/dataset. Multiple datasets are stored on a 500Mb hard disc contained in a PCMCIA card. A block diagram showing the salient features of the channel sounder is shown in *Figure 1*. The features can be summarised as

a) 256 bit sequence (modified PRB sequence) clocked at 256kHz (1kHz spectral lines)
b) Data capture: 65536 samples in 256ms (spectrum information at 3.9Hz intervals)
c) Signal averaging of 256 sample sets in 66s (total 16M samples, increases SNR by 24dB)
d) 65536-point FFT to obtain spectrum

For sounding purposes we do not need a particularly fine resolution so, initially, we are using a 256-chip sequence, clocked at around 256kHz. The spectral lines will therefore be 1kHz apart. The equipment is capable of generating much longer sequences than this. A 256-point FFT would be sufficient to extract meaningful data, but sampling in greater detail allows us to view the background noise and interfering signals at a high resolution. We will collect 2^{16} samples, taking 256ms to do so, and resulting in spectrum information every 3.9Hz. We will use signal averaging to increase the SNR, collecting 2^8 sample sets of 2^{16} samples in 66s. This increases the SNR by 24dB, and reduces the noise-equivalent bandwidth by 256 times from something of the order of 3.9Hz (depending on the FFT details) to 0.015Hz. This requires the receiver to maintain a very good synchronism to the transmitter, which is achieved using a code- or delay-locked which, in turn, requires attention to the sequence design to ensure that a good impulse response can be extracted at the receiver.

3. EQUIPMENT DESIGN

3.1 Transmitter Antenna

Without a careful analysis, it is not always clear whether, for inductive communications with portable equipment, an air-cored loop makes a better transmitter antenna than a ferrite-cored solenoid. But whichever type of antenna is used it will, unless physically very large, present a mainly inductive load to the driver. The power transfer efficiency of an analogue power amplifier drops by a factor of $Q = j\omega L/R$ when the load is inductive. We could envisage a spot-frequency system with some method of re-tuning the antenna between measurements, but this would probably require a cumbersome and time-consuming mechanical tuning system.

It seems clear that using an untuned wideband transmitter is the correct approach for our extremely wideband signal – a 16:1 frequency range, initially). It is desirable to drive such an antenna from a digital complementary output stage so that the issue of PA efficiency does not arise. However, it is still the case that, for a given driving voltage, the magnetic moment is lower, by a factor of Q, than it would be for a purely resistive loop. We have the choice of several methods of excitation – e.g. spot frequency, chirp, pseudo-random sequence and, of these, the wideband PRB sequence seems the most attractive.

As noted above, the magnetic moment is reduced by a factor of Q for single frequencies. However, driving the loop with a wideband NRZ signal allows us to utilise more power, owing to the presence of low frequency components in the signal. We have deduced, from a circuit simulation, that a 256 chip near-random sequence will allow us to drive the antenna with 256 times more power than if we used a simple 010101... sequence at the same chip rate. Even so, the amount of power we can send to the antenna is limited.

This sequence is integrated by the transmitter antenna to give a current waveform as shown in *Figure 1*, and is differentiated at the receiver antenna. The power spectrum for the integrated PRB sequence drops off smoothly at 6dB/octave for frequencies below the chip-rate. Since atmospheric noise also drops with frequency, this may turn out to be an appropriate spectrum to use.

3.2 Sequence Design

3.2.1 Choosing a Sequence

Using a maximal-length pseudo-random binary (PRB) sequence poses a problem, because binary M-sequences are of period 2^n-1 and thus cannot be used directly with the most efficient form of FFT, which requires a sequence of period 2^n. Two well-known solutions [5] to this problem are i) to ignore it, and to pretend that the continuous signal has a period of 2^n; and ii) to use the inverse Fourier transform to generate the test data itself. In the latter case, the resulting test signal will have the prescribed length and power spectrum, but will, unfortunately, consist of non-binary values whereas, for PA efficiency, we need a binary sequence, so that we can use a digital or class-D power amplifier.

The solution we have adopted is to take a maximal-length PRB sequence and add one bit, to make a sequence of length 2^n. The position of the added bit affects the size of the sidelobes in the auto-correlation function (ACF) but, by trial and error, it is possible to find the sequence with the lowest sidelobes although, as *Figure 2* shows, the sidelobes are still significant. The sequence is listed in *Table 1*.

Table 1. A unipolar M+1 sequence, formed by adding one bit (shown ringed) to a 255-bit binary M-sequence. Table reads across, then down.

```
10110001,11101000,01111111,10010000,
10010011,11101010,10111000,00110001,
01011001,10010111,11101111,00110111,
01110010,10100101,00010010,11010001,
10011100,11110001,10110000,10001011,
10101111,01101111,10000110,10011010,
11011010,10000010,01110110,01001001,
10000001,11010010,00111000,10000000.
```

This 'M+1' sequence has the advantage that, because it has no d.c. content, the induction loop (which acts as an integrator) will not cause the current to ramp over time. Clearly, the precise power spectrum of the sequence does not matter for simple spectral measurements of attenuation and phase shift. Although the choice of sequence is not crucial for spectral measurements, a sequence with a good auto-correlation (i.e. minimal side-lobes) is necessary for measurements of the channel's impulse response. A good system identification signal is also required for receiver synchronisation using the code-locked loop. Both these requirements are usually met by using a sequence with a good ACF, which we do not have here. It is, however, possible to use a non-ideal test sequence provided that the input-output cross-correlation is carried out between the applied sequence and its 'inverse', which we will now derive.

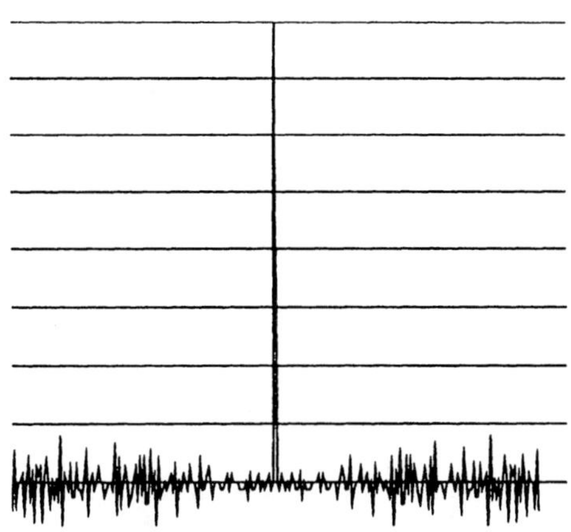

Figure 2. ACF of M+1 sequence (linear scale).
The ACF of the sequence in *Table 1* shows sidelobes, at up to –21dB relative to the peak. The r.m.s. value of the sidelobe function is –30dB. (The r.m.s. of the peak is -24dB [1/√256] of course, so the peak / sidelobe energy ratio is only 6dB). Y-axis is linear, with values 0–1.

3.2.2 Cross-correlation with the Inverse Sequence

Our transmitted test signal is a periodic sequence of N samples with sampling interval T, represented by $f(nT)$. We wish to correlate this with a sequence in the receiver, such that we obtain the ideal CCF, a delta function. We can therefore *define* the receiver's copy of the sequence as the *inverse*, $f^{-1}(nT)$, and note that we desire

$$f(nT) \oplus f^{-1}(nT) = \delta(nT) \tag{1}$$

where $\oplus$ denotes a normalised correlation and δ is unity at $nT = 0$, otherwise zero. If $f(nT)$ were, for example, a unipolar binary M-sequence (1,0,0,1,...) then we could note that its inverse would be a bipolar version of the same sequence (1,-1,-1,1...) that is suitably scaled.

Solving (1) for the inverse of f was described in [6, 7] but here we present a more straightforward approach. First, we apply the convolution theorem and re-arrange to give

$$F^{-1}(m\Omega) = \frac{N}{\tilde{F}(m\Omega)} \tag{2}$$

where F is the discrete Fourier transform (DFT) of f, defined such that $f(nT) = 1 \Rightarrow F(m\Omega) = N\,\delta$; and ~ denotes a reversal of the order of the coefficients of F due to the fact that the operation in (1) is correlation and not convolution. We note that because f contains only real coefficients, reversing the order of the coefficients of F is equivalent to using the complex conjugate, F^*. Thus, taking the inverse DFT (IDFT) of both sides, we have

$$f^{-1}(nT) = \textbf{IDFT}\left(\frac{N}{F^{*}(m\Omega)}\right) \tag{3}$$

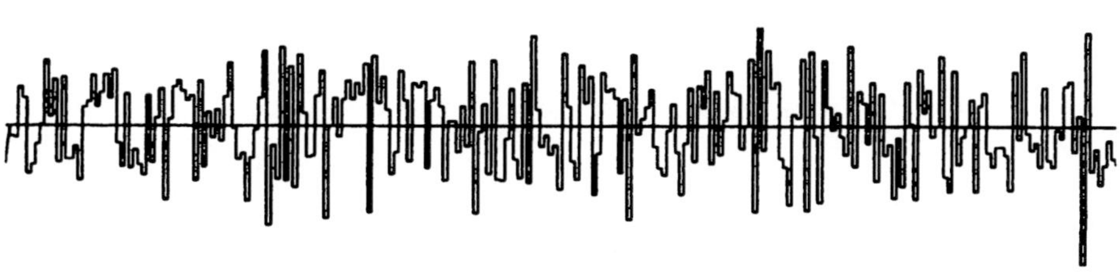

Figure 3. The inverse of the sequence of *Table 2*

Using this to calculate the inverse of the function in *Table 1* results in the function plotted in *Figure 3*, and listed in *Table 2*.

Table 2. The inverse of the unipolar M+1 sequence listed in *Table 1* and plotted in *Figure 3*. Table reads across, then down.

-0.59,	-0.64,	2.18,	1.55,	-2.61,	-2.11,	-1.00,	0.04,
3.62,	0.53,	2.58,	-2.03,	2.69,	-1.79,	-1.83,	-1.13,
-3.08,	1.01,	1.40,	2.82,	1.05,	1.88,	2.90,	1.57,
3.13,	-0.96,	-2.25,	1.79,	-2.31,	-1.31,	-2.10,	-2.75,
1.69,	-2.02,	0.42,	2.05,	-4.10,	0.39,	2.21,	2.47,
1.91,	1.47,	1.71,	-3.07,	2.53,	-2.25,	0.76,	-0.61,
0.89,	-0.80,	1.65,	3.48,	-0.13,	-1.82,	-1.49,	-4.20,
-1.75,	-0.17,	1.57,	4.11,	-5.48,	-1.04,	-3.00,	4.39,
-3.07,	3.34,	-3.43,	3.93,	0.70,	-1.70,	-1.63,	1.67,
3.06,	-5.11,	0.00,	1.59,	-0.57,	1.49,	2.52,	1.50,
2.39,	1.75,	3.45,	-4.81,	3.86,	1.23,	2.77,	0.87,
-2.65,	-1.46,	3.06,	1.46,	-1.92,	2.35,	2.18,	2.43,
-2.29,	1.42,	2.10,	1.09,	-3.08,	0.28,	0.22,	-1.43,
1.05,	-1.13,	3.54,	-4.84,	-0.31,	1.17,	-2.62,	3.65,
-2.97,	-3.08,	-0.71,	2.05,	-1.80,	-2.91,	2.37,	-3.19,
5.01,	0.85,	-1.12,	-0.50,	-1.53,	-1.05,	-3.55,	4.00,
1.04,	-1.96,	-3.00,	3.36,	1.25,	2.72,	-3.82,	-1.49,
2.91,	1.91,	2.06,	1.54,	-2.54,	1.49,	-5.22,	3.93,
-0.44,	0.35,	1.02,	1.98,	-1.14,	-2.32,	-2.76,	-0.24,
1.21,	-1.44,	-3.83,	-0.94,	2.13,	-1.94,	2.25,	1.00,
3.04,	-2.09,	1.95,	-1.59,	0.32,	3.00,	1.87,	1.77,
-1.24,	0.01,	3.66,	-4.73,	5.57,	-1.30,	4.17,	2.10,
2.45,	-1.73,	-2.22,	-4.35,	0.63,	0.44,	3.70,	-4.67,
4.16,	-0.59,	-4.27,	3.63,	1.01,	-0.19,	0.70,	-0.53,
-1.57,	4.47,	-2.20,	1.96,	1.26,	-0.91,	2.42,	-3.06,
1.81,	-1.56,	-0.99,	-4.01,	-0.65,	-3.38,	2.44,	0.12,
-4.09,	3.10,	0.75,	1.97,	-0.53,	0.03,	3.90,	-2.81,
-3.61,	3.07,	-2.07,	-0.56,	0.01,	1.52,	-3.62,	1.15,
1.88,	-1.41,	-2.18,	-1.11,	-1.09,	-1.56,	-3.55,	3.00,
-0.66,	4.12,	-0.35,	-1.29,	-0.86,	-2.22,	2.17,	-1.82,
-2.17,	-0.16,	-0.63,	2.22,	2.48,	-1.36,	0.53,	-7.71,
5.26,	-2.46,	-1.25,	-3.23,	-2.12,	-0.75,	-1.78,	-2.08.

Equation 3 is easily evaluated in a package such as MATLAB [8]. The salient line of the program, with vectors 'testFn' and 'InverseFn' is

```
InverseFn = ifft( N / conj(fft(testFn)) )
```
(4)

The cross-correlation of the unipolar sequence of *Table 1* with the inverse of *Table 2* gives a result (calculated using MATLAB) with sidelobes all below –63dB. The r.m.s. value of the sidelobe function is –74dB. With a higher-precision inverse the CCF comes close to an ideal delta function.

3.2.3 Observations on the Behaviour of the Inverse

There are several observations we can make. Firstly, the sum of the coefficients in *Table 2* is exactly 2; this arises because the product of the sum of the coefficients of f and its inverse, respectively, is N. This statement is proved by noting that the mean of the coefficients of f, given by $1 \oplus f$, represents a d.c. level equating, (because of the DFT definition we are using) to $F(0) / N$ so the sum of the coefficients of f is $F(0)$ and thus we have a restatement of (2) for the special case of $m\Omega = 0$.

A second observation, from (3), is that the inverse does not exist if $F(m\Omega)$ contains one or more zeroes. One case where F contains a zero is where the test signal is required to have no d.c. component – as in the system under discussion here. To resolve the situation we use $f(nT)$ as the test signal, which we use to calculate the inverse function, but we note that because the d.c. component is not present at the receiver, the correlation is

$$\left\{f(nT)-\frac{F(0)}{N}\right\}\oplus f^{-1}(nT)=\delta-\frac{F(0)}{N}\oplus f^{-1}(nT) \tag{5}$$

and thus the effect is simply to add a constant to the output. From the observation on the sum of the coefficients, we can see that (5) is simply

$$\delta-1/N \tag{6}$$

which is therefore the cross-correlation product when we use a copy of f that has been level-shifted so that the mean of the coefficients is zero. The d.c. level is therefore $-1/(N-1)$ times the peak – see *Figure 4*.

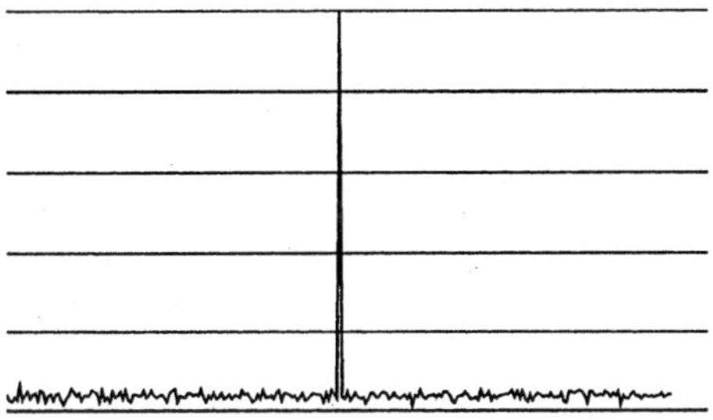

Figure 4. CCF of bipolar M+1 sequence with the unipolar inverse (Y-axis 10dB/division). The near-ideal CCF of *Table 1* with *Table 2* has very little sidelobe energy. If, instead, the correlation is performed with a bipolar version of the M+1 sequence, then the sidelobes have a near-constant level of 1/255 of the peak. (r.m.s. of sidelobe function is –48dB).

It is interesting to look at the situation where, in addition to removing the d.c. component of f, we shift the d.c. level of the inverse function by an arbitrary amount, k. Now an extra term appears on the right of 5), of

$$k\oplus\left\{f(nT)-\frac{F(0)}{N}\right\} \tag{7}$$

but this is proportional to the mean of the expression in brackets, which is zero by definition. Hence we can deduce that, with no d.c. component to f, the d.c. level of the inverse f^{-1} does not affect the correlation product. We can turn this around and say that if the d.c. component of the inverse is removed then the d.c. level of the incoming sequence does not matter.

3.2.4 Noise Factor

It is apparent that the cross-correlation we are undertaking is similar to that of a matched filter and so we should expect the noise factor to depend on the nature of the sequence. Following a derivation similar to that for a matched filter, we can show [9] that the noise factor is given by

$$1 \Big/ \frac{1}{N}\sum_{n=0}^{N-1}\left|f(nT)\right|^2 \cdot \frac{1}{N}\sum_{n=0}^{N-1}\left|f^{-1}(nT)\right|^2 \tag{8}$$

This has a maximum value of unity and expresses the degradation in SNR caused by using an unmatched system. As with a matched filter, the ideal SNR occurs when the sequence is auto-correlated, that is when $f = f^{-1}$.

Using, as an example, the unipolar M+1 sequence and its inverse, the noise factor is calculated to be 0.33, or –3.6dB. This loss can mostly be attributed to the fact that half of the bits in the transmitted sequence contain no energy. A calculation shows that if, instead, we transmit a bipolar copy of the sequence the noise factor is 0.656 or –1.83dB and the cross-correlation product is the expected δ - 1/256. We can further maximise the noise factor by a small amount by using a copy of f^{-1} with no d.c. component. It was noted earlier that if the d.c. level of f were zero then the d.c. level of f^{-1} would not affect the cross-correlation product, but the noise factor is improved marginally by this action. The reason for this is that for any set of integers a, the sum $\Sigma(a - m)^2$ is a minimum when the constant m is the mean of the set, i.e. $m = \Sigma a/N$ or equivalently $\Sigma(a - m) = 0$.

3.3 Code-locked Loop

3.3.1 Conventional early-minus-late operation

Because we are dealing with very low signal levels we are using a form of signal averaging that requires very long sample times, of the order of 30–60s. For the receiver's sample frequency to remain locked to the transmitter over this length of time we require to extract a synchronisation signal from the sounding sequence, by cross-correlating it with its inverse in the receiver. A standard method uses a code-locked loop (CLL), also known as a delay-locked loop, as shown in *Figure 5*. This requires two correlations – one with an 'early' copy of the inverse sequence and one with a 'late' copy; with a relative time-shift of one 'chip' between the two copies. The two CCFs are differenced and the result is a bipolar error signal that is used to control the master clock frequency to bring the system into lock, in a similar way to that of a conventional phase-locked loop.

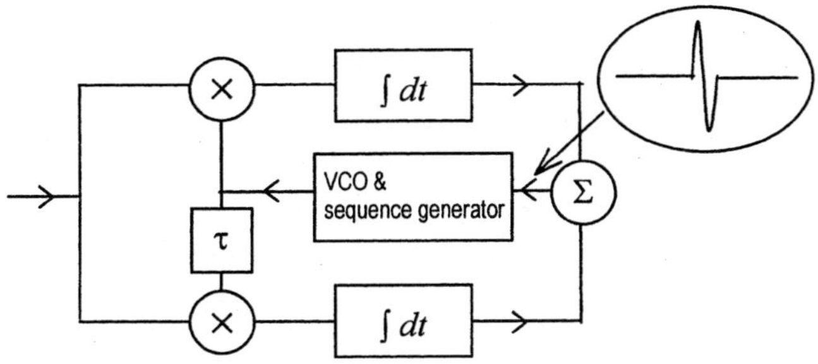

Figure 5. Operation of a conventional code- or delay-locked loop.

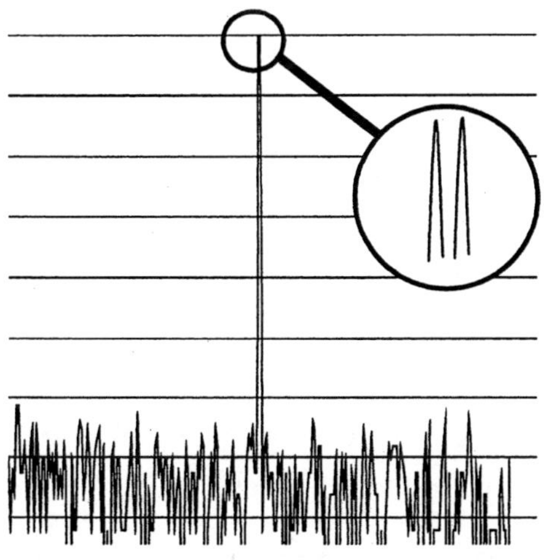

Figure 6. The calculated sum of Early–Late CCFs (Y-axis shows power spectrum at 10dB/division). The d.c. level in the sidelobes cancels virtually completely leaving just noise due to rounding errors in the calculation of the inverse sequence.

Equation (5) shows that the correlations result in a d.c. level, but we know from the near-ideal CCF of the unipolar function and its calculated inverse that there is very little ripple in the sidelobes, so the action of taking the difference between the 'early' and 'late' correlations must be to cause the d.c. level to almost cancel, providing us with the bipolar control signal (shown inset in *Figure 5*) that we require for the code-locked loop. The calculated *early–late* sum is shown, as a power spectrum in *Figure 6.*

3.3.2 A system control signal for decreased acquisition time

It is clear that the *early–late* correlation could be performed in a single operation by combining the reference sequences into one – call it h – with

$$f(nT) \oplus h(nT) = r(nT) \tag{9}$$

where r is the desired system control signal, and is traditionally the 's-curve' shown inset in *Figure 5*. But we are now able to define r to be *any* desired function, and to calculate h appropriately. For example, if r is the function shown in *Figure 7* this may lead to a decreased acquisition time, which may be important at the low frequencies, we are investigating with the channel sounder

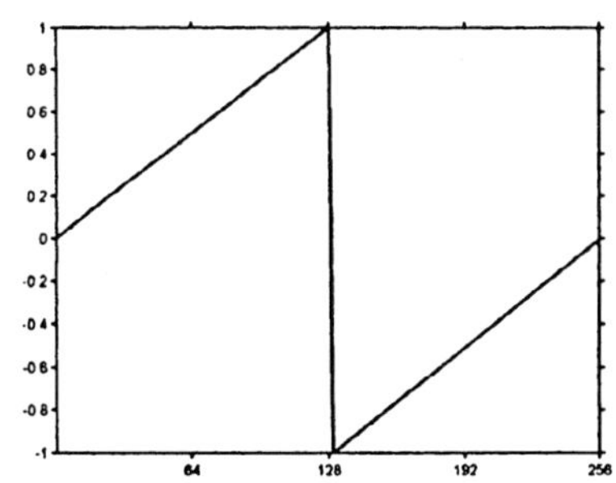

Figure 7. This control signal may decrease acquisition time at the expense of noise factor.

For a code-locked loop the appropriate figure of merit is a noise-equivalent bandwidth. This is not as easy to calculate as the noise factor discussed above. Nevertheless, without recourse to a full analysis, we can still use (8) as a 'figure of merit' for the system. Based on this, a calculation shows that the extended-range control signal is far from ideal in terms of noise. This can be attributed to the high r.m.s. value of the sequence. Clearly, if we were to use such a sequence for system acquisition, it would pay to switch to the traditional s-curve function once lock was obtained.

3.4 Receiver Antenna

The design of the receiver front-end owes more to low-noise audio design than to r.f. techniques; impedance matching, for example, is not an issue. However, in order to obtain the best SNR, noise-matching is important. This aspect of the design was outlined in [1].

4. PRELIMINARY RESULTS

4.1 Equipment and Data Processing

Initial tests with the receiver used a slightly different specification to that given in §2.3. Data was processed at 500kHz, capturing 2^{17} samples per dataset and resulting in spectral data at 3.8Hz intervals. The 16-bit analogue to digital converter (ADC) was an AD7721 from Analog Devices, featuring an over-sampling anti-aliassing digital filter with a bandwidth of 244kHz.

The data was processed using a straightforward 131072-point discrete Fourier Transform. No windowing function was applied, although windowing will be undertaken in the future as an aid to the exact characterization of the sidebands of interfering stations.

For these preliminary results, the receiver antenna was not calibrated. It featured a low-pass filter to remove some of the mains interference, and a high-pass filter to aid the anti-aliassing provided by the ADC. It was believed to have a substantially flat response between those limits.

4.2 Self-Noise and Signal Averaging

Operating the prototype channel sounder receiver with a shorted input showed a broadband noise with a mean spectral density of 3.5μV/√Hz), which was higher than expected. Applying signal-averaging over 100 datasets reduced this a factor of almost ten, as expected. However, several noise spikes were not affected by the averaging, thus demonstrating that they arose from a synchronized source (*Figure 8*). The spikes are at exact sub-multiples of the clock frequency of 500kHz. The largest spike is 80μV at 125kHz. This has implications for sounding, and future work will have to look at ways of reducing this synchronized self-noise.

4.3 Spectrum Survey

The receiver section of the channel sounder has been used to undertake a preliminary spectrum survey. *Figure 9*, is a typical laboratory result showing the horizontal field with the antenna directed towards the BBC Radio 4 transmitter (198kHz, 500kW) at Droitwich, some 180km from Leeds, bearing 197° magnetic. The test antenna had only a low gain, so the sidebands of this signal were partially buried in noise. Nevertheless, a clear 198kHz carrier could be observed. (This is difficult to see at the scale of the printed Figure). The 60kHz carrier of the NPL's Standard Time transmission from Rugby was also observed.

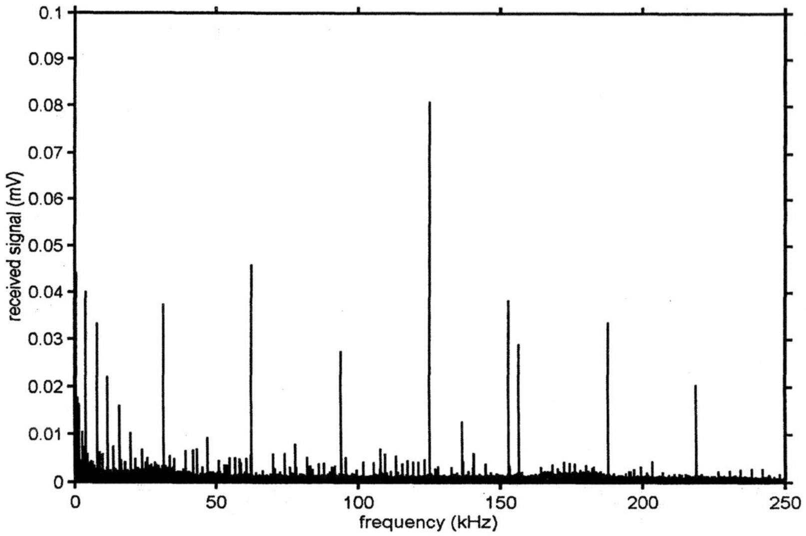

Figure 8. The self-noise of the channel sounder after averaging 100 datasets. The spikes represent internal interference. (Lines have been thickened for clarity).

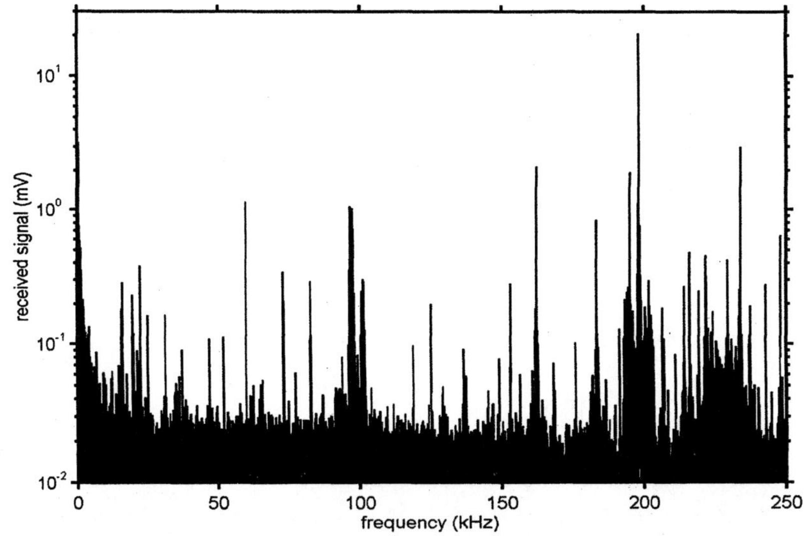

Figure 9. A typical laboratory reading of horizontal field. Note the 60kHz time signal carrier, broadband nav. beacon at 100kHz and BBC R4 at 198kHz. (Lines thickened for clarity).

A number of other stations are apparent but the largest, by far, is the 2kHz-wide signal at 110kHz, magnitude 22mV, which is assumed to be one of the Loran navigation beacons, although this has yet to be confirmed.

There is a strong signal at 15.625kHz and its harmonics. These are sub-multiples of 500kHz but, unlike the self-noise signal mentioned above, the spectrum of these lines is slightly smeared thus demonstrating that they must originate (for the most part anyway) from an unsynchronized signal; this being the TV timebase for the UK's 625-line PAL television system.

Below a few kHz, individual mains harmonics are evident (although not at the scale of this plot). The test antenna did not fully filter the 50Hz mains interference which was one reason why the gain of the antenna had to be kept low.

It was noticeable that unless all nearby computer equipment was turned off there was considerable additional interference present during the lab. readings; notably timebase signals at harmonics of 48.4kHz from high-resolution TV monitors.

4.4 Channel Sounding

The next stage of the project will be to undertake a full sounding exercise by transmitting a synchronized wideband sequence. The data-logging equipment will be programmed to take automated measurements at a remote underground location over a period of several months. This work had to be postponed in 2001 because the outbreak of Foot and Mouth Disease in the UK made access to caves impossible for most of the year.

It is known that atmospheric noise varies with the seasons, and we expect that there might also be a variation in underground propagation, due to the changing water content of the peat overburden in many limestone areas of the UK. Overall, we expect various effects to contribute towards there being an optimum frequency for transmission, which varies with conditions, as discussed in [1].

5. CONCLUDING REMARKS

This paper has briefly described the design of a wideband channel sounder, which can also be used for noise and interference measurements in the LF band. The salient points of the design are as follows.

a) The extreme wideband nature of the sounding signal requires a careful approach to the antenna design and includes the necessity to transmit a binary (two level) sequence.
b) For ease of processing this must be of length 2^n but such a sequence has a poor auto-correlation property.
c) The low signal levels imply a long sampling time, and so good synchronisation between the receiver and transmitter is important.

d) This is achieved by calculating an inverse to the transmitted sequence and cross-correlating to obtain the impulse response of the system.
e) The transmitted signal remains bi-valued for efficiency, but the inverse sequence is multi-valued.
f) The transmitted sequence must contain no zeroes in the frequency domain in order that an inverse can be calculated. The d.c. component of the inverse sequence is removed, which introduces an offset in the correlation product at the receiver, but renders it insensitive to the d.c. level of the received signal, which will largely be absent.
g) Our approach to sequence design may lead to a simple method of reducing the acquisition lime of the code-locked loop at the very low frequencies we are using.

This sounder is part of an adaptive communications system already described [1]. The next phase of the research programme will be to validate our propagation model via practical sounding measurements and to collect data on noise and interference in the LF band. A complete adaptive data transmission system architecture will then be tested.

REFERENCES

[1] D. Gibson & M. Darnell, "Adaptive digital communications for sub-surface radio paths," *Proc. 5th International Symposium on Digital Signal Processing for Communications Systems* (*DSPCS '99*), Perth, Australia, 1st-4th Feb. 1999. pp237-244.

[2] D. Gibson, "A channel sounder for sub-surface communications: part 2 – Computer simulation of a small buried loop," *BCRA Cave Radio & Electronics Group Journal* **41**, pp29-32, Sept 2000. ISSN 1361-4800, **http://caves.org.uk/radio/**

[3] M.G. Laflin, "ELF/VLF spectrum measurements," AGARD-CP-529, pp17-1 to 17-12. (*Conference Proceedings **529** on ELF/VLF/LF Radio Propagation and Systems Aspects; Advisory Group for Aerospace Research and Development; NATO*).

[4] D. Gibson, "LF utility stations", *BCRA CREGJ* **28**, pp16-17, June 97, ISSN 1361-4800, **http://caves.org.uk/radio/**.

[5] P.E. Wellstead, "Pseudonoise test signals and the fast Fourier transform," *IEE Electronics Letters*, **11**(10), 15 May 1975

[6] A. Al-Dabbagh, M. Darnell, A. Noble, and S. Farquhar, "Accurate system identification using inputs with imperfect autocorrelation properties," *IEE Electronics Letters* **33**(17), pp 1450-1451, 14th August 1997.

[7] A. Al-Dabbagh, and M. Darnell, "A novel channel estimation technique for radio communications", *DSPCS '99*, pp245-251.

[8] MATLAB technical computing software is described at **http://www.mathworks.com**

[9] D. Gibson, *Sequence Design for System Identification*, **http://caves.org.uk/radio/**

Chapter 10

COMPUTATIONAL COMPLEXITY OF ITERATIVE CHANNEL ESTIMATION AND DECODING ALGORITHMS FOR GSM RECEIVERS

Hailong Cui and Predrag B. Rapajic

School of Electrical Engineering and Telecommunications, the Universtiy of New South Wales, Sydney, Australia

Abstract: In this chapter we investigate the balance between complexity and BER performance of 4 different types of GSM receiver: The conventional training sequence based channel estimation receiver, iterative channel estimation feedback from equalizer, iterative channel estimation plus iterative equalization and decoding, and combination of iterative channel estimation and Iterative equalization and decoding. The combination of iterative channel estimation and iterative equalization and decoding has the best trade off among the receivers with iterative processes. It has about 1.3 dB gain over conventional receiver using training sequence estimation only.

Key words: iterative channel estimation, equalization, decoding

1. INTRODUCTION

Intersymbol Interference (ISI) caused by multi-path propagation is a common problem in digital mobile communication system. In addition, the channel is time varying due to the movement of the mobile station relative to its surrounding. In a time division multiplex access (TDMA) system like GSM, the receiver must be able to estimate the channel and compensate for channel distortion adaptively. The performance of this compensation is largely determined by the accuracy of the channel estimation. In a conventional GSM receiver, the channel impulse response (CIR) is estimated using the known training sequence transmitted as a midamble in each burst.

Feeding back the estimated data output from the equalizer or decoder as an extended training sequence can improve the accuracy of the estimated CIR. Which is called iterative channel estimation (ICE) [1].

The channel encoder and ISI channel can be treated as a serially concatenated coding scheme, the decoding principle of Turbo Code [2] can be used to perform iterative equalization and decoding [3]. Four different types of GSM receiver with these techniques were introduced. The complexity of implemented these receivers and the corresponding performances are compared here in terms of number of operations required in receiving one GSM block.

This chapter is organized as follows: in section 2, we introduce the configuration of four candidate GSM receivers employing various iterative schemes. Their complexity analysis is presented in section 3. Simulation results of bit error rate (BER) performance are presented in section 4, and the conclusion is made in section 5.

2. CANDIDATE GSM RECEIVERS

The typical structure of the GSM digital receiver is shown in Figure 1. The channel impulse response (CIR) is estimated by a known training sequence or data sequence transmitted as a midamble in each GSM burst. The channel is estimated using channel sounding (CS) [4] or least square (LS) [5] techniques, and this estimated CIR is used in equalization process to eliminate the impact of Intersymbol Interference (ISI). After de-interleaving, the decoder estimates the original message sequence. In this paper, we consider receivers which employ maximum a posteriori probability (MAP) [6] algorithm for equalization and either maximum likelihood sequence estimation (MLSE) [7] or MAP estimation in the decoder. The MAP algorithm is a soft-in soft-out (SISO) algorithm, which is well suited to iterative receiver schemes.

2.1 Conventional Training Sequence Receiver (TS)

In conventional training sequence receiver, the channel impulse response was estimated by the known training sequence stored at the receiver. MLSE or MAP is used to perform equalization and MLSE implemented by Viterbi algorithm to perform decoding. The output from the MAP equalizer will be soft value, which can improve the performance of decoder.

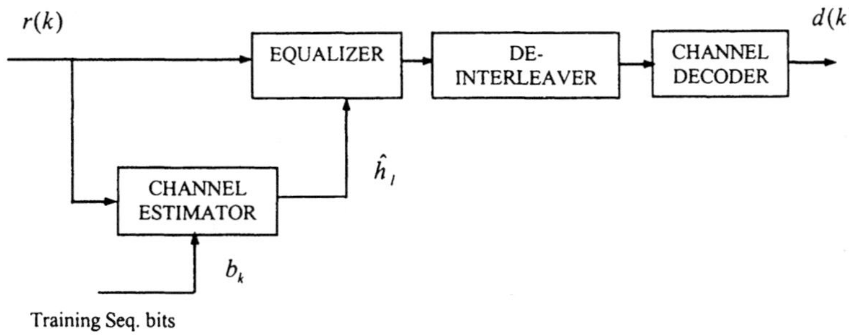

Figure 1. Conventional GSM Receiver [1]

2.2 Iterative Channel Estimation from Equalizer (EQ-ICE)

The effectiveness of the Equalization depends on the accuracy of the reliability of the estimated CIR. In order to get more accurately estimation, the equalization output can be feedback to the channel estimator to perform Iterative Channel Estimation (ICE), the performance is further improved using soft decision feedback [8] This configuration is shown in Figure 2.

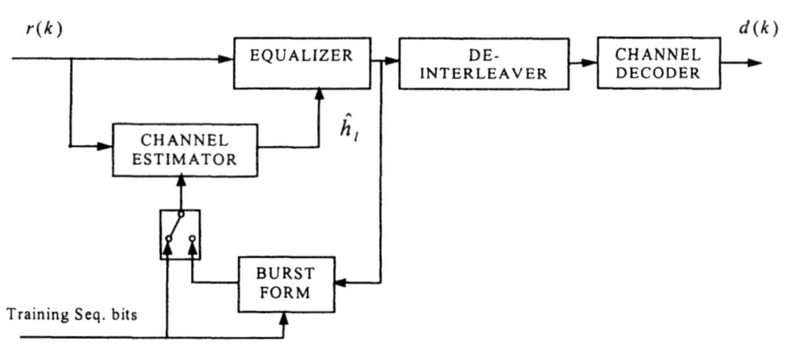

Figure 2. Iterative channel estimation feedback from equalizer [1]

2.3 Iterative Channel Estimation plus Iterative Equalization and Decoding (EQ-ICE+Turbo)

In GSM, the channel encoder and ISI channel can be viewed as a serially concatenated coding scheme, so the decoding principle of Turbo Code can be used to perform Iterative Equalization and Decoding [3]. So in this configuration there are two iterative processes: One for channel estimation and the other for equalization and decoding.

2.4 Turbo Iterative Channel Estimation (Turbo-ICE)

These two iterative processes can be combined to perform Turbo Iterative Channel Estimation. The output from the APP decoder were fed back to the channel estimator to re-estimate the channel, using the improved CIR and the extrinsic information as a priori knowledge, the APP equalizer will get improved data estimation.

MAP algorithm is used to perform these iterative processes. In order to reduce the complexity, a simplification of MAP algorithm: Max-Log-MAP [9] algorithm is used instead MAP in these receiver schemes. All the iterative processes can be performed any times without delay restriction [10]. One iterative process is performed for each in our comparison.

3. COMPLEXITY COMPARISION

The complexity is evaluated by the total operation needed in receiving one GSM block. The interleaving and de-interleaving processes are ignored in this complexity calculation. Channel sounding is used as channel estimation algorithm for both training sequence estimation and iterative channel estimation. The operation needed for one GSM burst is shown in Table 1. For soft decision feedback [8], a look up table used to convert the log likelihood ration (LLR) of each transmitted data into soft values.

Equalization is performed burst by burst and the decoding is taken when it receives the whole speech block. So the equalization and decoding complexity are estimated in term of the number of operation needed for equalizing one burst and decoding one speech block. Table 2 gives the complexity of Viterbi equalizer and decoder used in conventional training sequence receiver.

Table 1. The Complexity of Channel Sounding

Operation	Channel Sounding		
	Training Sequence Estimation	Iterative Channel Estimation	
		Hard Decision Feedback	Soft Decision Feedback
Multiplication	$M_{CIR}L_t$	$M_{CIR}(L_B - M_{CIR})$	$M_{CIR}(L_B - M_{CIR})$
Addition	$M_{CIR}(L_t - 1)$	$M_{CIR}(L_B - M_{CIR} - 1)$	$M_{CIR}(L_B - M_{CIR} - 1)$
Look-ups			$L_B - M_{CIR} - 26$

M_{CIR}: Number of estimated channel impulse response taps.

L_t : Number of training sequence bits used in channel estimation.

L_B: Length of a GSM normal burst excludes guard bits.

Table 2. Complexity of Viterbi Equalization and Decoding

Operation	Viterbi Equalizer for GSM	Viterbi Decoder for GSM
Addition	$(M+2)\ 2^M (L_B + M)$	$8\ 2^K (L_{C1} + K)$
Multiplication	$(M+1)\ 2^M (L_B + M)$	$4\ 2^K (L_{C1} + K)$
Max Operation	$2^M\ (L_B + M)$	$2^K\ (L_{C1} + K)$

M : Memory Length of the ISI channel $M = M_{CIR} - 1$

L_B : Number of bits in one GSM normal burst excludes guard bits.

K : Memory length of the channel encoder for GSM.

L_{C1} : Length of Class1a bits in one GSM speech block.

Max-Log-Map is used to perform iterative channel estimation, equalization and decoding. The operation required for equalizer and decoder is presented in Table 3.

Table 3. Complexity of Max-Log-MAP Equalization and Decoder

Operation	Max-Log-MAP Equalizer				
	Branch Transition Probability γ Calculation	Forward Process α Calculation	Backward Process β Calculation	Log Likelihoo d Function	Total (Approximation)
Addition	$2\cdot(M+3)\cdot 2^M\cdot(L_B+M)$	$2\cdot 2^M\cdot(L_B+M)$	$2\cdot 2^M\cdot(L_B+M)$	$5\cdot 2^M\cdot L_B$	$(2M^2+10M+10L_B+2ML_B)\cdot 2^M$
Multiplication	$2\cdot(M+3)\cdot 2^M\cdot(L_B+M)$				$(2M+6)(L_B+M)\cdot 2^M$
Max Operation		$2^M\cdot(L_B+M)$	$2^M\cdot(L_B+M)$	$2\cdot 2^M\cdot L_B$	$(4L_B+2M)\cdot 2^M$

L_B : Number of bits in one GSM normal burst excludes guard bits.

M : Memory Length of the ISI channel. $M = M_{CIR} - 1$

Operation	Max-Log-MAP Decoder					
	Branch Transition Probability γ Calculation	Forward Process α Calculation	Backward Process β Calculation	Log Likelihood Function (Info Bits)	Log Likelihood Function (Coded Bits)	Total (Approxi mation)
Addition	$4\cdot 2^K\cdot(L_{C+1}+K)$	$2\cdot 2^K\cdot(L_{C1}+K)$	$2\cdot 2^K\cdot(L_{C1}+K)$	$5\cdot 2^K\cdot L_{C1}$	$2\cdot 2^K\cdot L_{C1}$	$(13L_{C1}+6K)\cdot 2^K$
Multiplicatio n	$6\cdot 2^K\cdot(L_{C1}+K)$					$6(L_{C1}+K)\cdot 2^K$
Max Operation		$2^K\cdot(L_{C1}+K)$	$2^K\cdot(L_{C1}+K)$	$2\cdot 2^K\cdot L_{C1}$	$4\cdot 2^K\cdot L_{C1}$	$(8L_{C1}+2K)\cdot 2^K$

K : Memory Length of Channel Encoder for GSM

L_{C1} : Length of Class1a bits in one GSM speech block

The number of process times required in receiving one GSM speech block is shown in Table 4. One iteration is investigated for each configuration because the simulation result in [3][12] show that most of the improvement is achieved by the first iteration. The GSM speech block are divided into 8 sub-blocks and interleaved into 8 GSM burst, only receiving the first speech block required equalization of 8 burst. For the other speech block, 4 sub-block data have been equalized in the previous 4 bursts due to the interleaving process [11]. In our simulation:

. Estimated channel impulse response taps: $M_{CIR}=5$

. Memory Length of the ISI channel $M = M_{CIR} - 1 = 4$

. Channel Encoder Memory of GSM: $K = 4$
. Length of one GSM Normal Burst: $L_B = 148$
. Length of training sequence used in channel estimation: $L_t = 16$
. Length of coded Class 1 bits in one speech block: $L_{C1} = 378$

Table 4. Configuration of Different Receiver Schemes

RECEIVER TYPE	Number of Operation in Receiving one GSM Speech Block					
	TS Channel Estimation	Iterative Channel Estimation	Equalization		Decoding	
			Viterbi	Max-Log MAP	Viterbi	Max-Log MAP
TS	4	0		4	1	
EQ-ICE	4	4		8		1
EQ-ICE+Turbo (N iteration)	4	4		4(2+N)		1+N
Turbo-ICE (N-iteration)	4	4N		4(1+N)		1+N

TS: Training Sequence Only Receiver using Max-log-MAP equalization.
EQ-ICE: Iterative Channel Estimation Feedback from Equalizer.
EQ-ICE + Turbo: Iterative Channel Estimation Feedback from Equalization plus Iterative Equalization and Decoding.
Turbo-ICE: Iterative Channel Estimation Feedback from Decoder with Iterative Equalization and Decoding.

Table 5. Complexity of Different Receiver Schemes

Operation	RECEIVER TYPE						
	TS	EQ-ICE		EQ-ICE+TURBO (1 Iteration)		TURBO-ICE (1 Iteration)	
		Hard Decision	Soft Decision	Hard Decision	Soft Decision	Hard Decision	Soft Decision
Addition	238652	432536	432536	686468	686468	511364	511364
Multiplication	160960	312236	312236	485100	485100	426732	426732
Max Operation	44512	125312	125312	212224	212224	173824	173824
Look-ups			976		976		976

In Table 5 we give the exact number of operation required for each type of GSM receiver in receiving one GSM speech block. Max-log-MAP is used in conventional GSM receiver to perform equalization.

4. SIMULATION RESULT

The simulation was done over EQ50 channel model provided by ETSI [11], it is an equalization test channel with vehicle moving at a speed of 50 km/h. The result shown in Figure 3 is using soft decision feedback for iterative channel estimation. From the simulation result of [8], we know that soft decision feedback has about 0.15 dB gain than hard decision only at a cost of some extra look-ups. Compared with EQ-ICE+Turbo in 2.3, Turbo-ICE has better performance and less cost. It has about 0.4-dB gain than EQ-ICE and a little more complexity. Without doubt, it is the best one among these 3 configurations (EQ-ICE, EQ-ICE+Turbo, and Turbo-ICE) when we trade off the complexity and performance. Compared with conventional training sequence receiver, Turbo-ICE has 1.3 dB gain and about 3 times complexity in implementation.

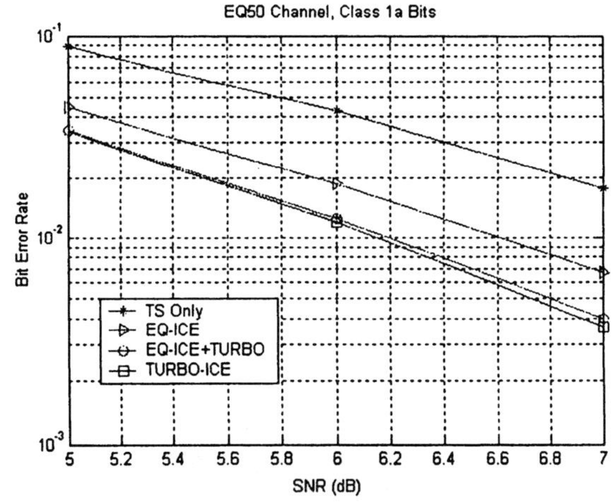

Figure 3. Performance of Different Receiver Schemes

5. CONCLUSION

In this chapter we investigated four different GSM receiver schemes and their complexity in implementation. Three kind of iterative processes have been studied: iterative channel estimation, iterative equalization and decoding, and combination of iterative channel estimation, equalization and decoding. Among these 3 configurations, the combination of iterative channel estimation, equalization and decoding has the best performance and complexity trade off. Using Channel Sounding channel estimation with soft decision feedback, it has about 1.3 dB gain over conventional training sequence receiver for class 1a bits in GSM. Similar improvement was also found for class 1b and class 2 bits.

ACKNOWLEDGEMENT

The authors wish to say thanks to Dr. Chris Nicol, Dr. Linda Davis, Dr. Graeme Woodward and Dr. Bing Xu for their support in this work.

REFERENCES

[1] Carlo Luschi, Magnus Sandell, Paul Strauch, Ran-Hong Yan, "Iterative Channel Estimation, Equalization and Decoding," *Internal paper of Lucent Technologies*, BL01131L-990225-04TM, 1999.

[2] C. Berrou, A. Glavieux, and P. Thitimajshima, " Near Shannon limit error-correcting and decoding: Turbo-codes (1)," *IEEE International Conference on Communications (ICC'93)*, pp.1064-1070, May1993.

[3] G. Bauch, V. Franz, " Iterative Equalization and Decoding for the GSM-system," *IEEE Vehicular Technology Conference, VTC 1999 -Fall*, Volume: 5, pp.2954-2958, 1999.

[4] G.L. Turin, " Introduction to spread-spectrum anti-multipath techniques and their application to urban digital radio," *Proceeding IEEE*, vol.68, pp.328-353, March 1980.

[5] L. Scharf, *Statistical signal processing: Detection, estimation and time series analysis*, Addison-Wesley, 1991.

[6] L. Bahl, J. Cocke, F. Jelinek, and J. Raviv, " Optimal Decoding of Linear Codes for Minimizing Symbol Error Rate", *IEEE Trans. Inform. Theory*, vol. IT-20, pp284-287. Mar 1974.

[7] G. D. Forney, Jr., "Maximum Likelihood Sequence Estimation of Digital Sequences in the Presence of Intersymbol Interference", *IEEE Trans. Inform. Theory*, vol. IT-18, pp. 363-378, May 1972.

[8] Sandell, M.; Luschi, C.; Strauch, P.; Yan, R, "Iterative Channel Estimation Using Soft Decision Feedback," *IEEE Global Telecommunications Conference, 1998. GLOBECOM 1998*, , Vol. 6, 1998, pp. 3728 –3733.

[9] P. Roberson, E. Villebrun, and P. Hoeher, " A Comparison of Optimum and Sub-Optimum MAP Decoding Algorithms Operating in the Log Domain," in *Proc. ICC '95*, June 1995, pp. 1009-1013.

[10] ETSI, GSM Recommendation 03.05, Digital cellular telecommunication system; Technical performance objectives, August 1996.

[11] ETSI, "Digital cellular telecommunications system; Radio transmission and reception," GSM 05.05 version8.4.0, 1999.

[12] Petermann, T.; Kuhn, V.; Kammeyer, K.-D. "Iterative blind and non-blind channel estimation in GSM receivers," *The 11th IEEE International Symposium on Personal, Indoor and Mobile Radio Communications, 2000. PIMRC 2000*, Vol. 1, 2000, pp. 554-559.

Chapter 11

MODELING AND PREDICTION OF WIRELESS CHANNEL QUALITY

Song Ci and Hamid Sharif
University of Nebraska-Lincoln

Abstract: In this chapter, we present an optimal frame size predictor for maximizing the wireless channel utilization. Kalman filter is adopted to predict the optimal frame size under noisy wireless channel. Simulations results show that the performance of proposed frame size predictor is much better than other prediction methods like moving average.

Key words: Wireless LAN, IEEE 802.11, Adaptive, Prediction, Indoor Wireless Channel

1. INTRODUCTION

Wireless channels are time varying and bursty due to affects of fading, noise, interference, shadowing, path loss, etc. This makes it very difficult to provide Quality of Service (QoS) for wireless data services. In order to combat these affects for the wireless channel quality, different approaches are proposed in literature. The link adaptation scheme is one approach focusing on providing the requested QoS through building reliable MAC and PHY layers. There are many research works centering on this topic. In [3, 2], link adaptation techniques are proposed through adapting link layer parameters such as frame size and fragment size. In [7], link adaptation techniques are discussed in terms of frame size and power control. In [5, 6, 9], adaptive modulation schemes are proposed for maximizing the channel efficiency and their performance are simulated and analyzed. In [8], an adaptive algorithm for optimizing the packet size used in wireless ARQ protocols is proposed. Two optimal packet size predictors are also proposed by using the maximum efficiency method and the maximum likelihood approach. However, that work is not based on the wireless LAN, which has several special concerns such as random exponential back off, deference, etc. So it is not appropriate for the IEEE 802.11 wireless LAN. In [4], an optimal frame size predictor based on the Kalman filter is proposed. However, there is no consideration of the random back off algorithm and the transmission overhead of MAC and PHY layers.

In this work, we present a link adaptation approach by pursuing Kalman filter to predict the optimal frame size. Initially, we analyze the network performance of the IEEE 802.11 wireless LAN to obtain the equation for the good throughput performance of the wireless LAN. Due to changes of the wireless channel quality as well as the network load such as the number of users inside a same service set, the good throughput performance is highly dynamic.

We propose to use the Kalman filter to predict the optimal frame size for the next transmission with respect to the maximum channel utilization. In order to analyze the proposed prediction algorithm, we conduct simulations with a field indoor channel measurement. The simulation results show that the proposed algorithm can substantially lower the estimation error by the order of tens compared with the traditional moving average method.

The rest of this paper is organized as follows. A brief review of current IEEE 802.11 wireless LAN standard is given in Section 2. An optimal frame size predictor based on the Kalman filter is introduced in Section 3. A simulation is described and the results are analyzed in Section 4. Finally, the paper is concluded with a conclusion.

2. REVIEW OF THE IEEE 802.11 WIRELESS LAN

The IEEE 802.11 wireless LAN is one of the most popular wireless LAN standards, which can provide functionalities that support a current style IEEE 802 LAN over wireless medium. Currently, the IEEE 802.11 wireless LAN standard is composed with two parts – MAC layer specifications and PHY layer specifications. There are two specifications in the PHY layer standard, which are the IEEE 802.11a and the IEEE 802.11b. The IEEE 802.11a adopts the OFDM modulation scheme that can provide up to 54 Mbps raw data rate while the IEEE 802.11b adopts the High rate DSSS system that can provide up to 11 Mbps raw data rate. At the time of this writing, the IEEE 802.11b has been widely deployed in the market. So in this work, we will focus on the throughput improvement of the IEEE 802.11b wireless LAN.

Currently, the IEEE 802.11a and the IEEE 802.11b, both are using the same MAC layer. There are two access control methods specified in the current standard, which are Distributed Coordination Function (DCF) and

Point Coordination Function (PCF). DCF is also known as CSMA/CA, which is an improved version of CSMA/CD used in Ethernet. PCF is actually a polling based access control method, which can theoretically eliminate collisions inside a same service area. Unfortunately, this method introduces large overhead and it has a very strict requirement of timing. Due to this, only limited devices with the PCF specification exist in market. Therefore, we only emphasize on the context of DCF.

DCF adopts both physical carrier sensing method and virtual carrier sensing method. Due to its CSMA/TDD nature and only one radio, a terminal of the wireless LAN cannot transmit and receive simultaneously. This makes it impossible to extend the traditional CSMA/CD method into the wireless domain since the prerequisite of CSMA/CD is that a terminal is able to receive and transmit data at the same time. Thus, the virtual sensing method is proposed to overcome this problem. The essence of the virtual sensing method is to utilize a timer, which records all timing information in the header of each frame received from other users. This virtual timer only updates its value when the value of the received timing information is larger than itself. An end user can transmit its frames only when both virtual carrier sensing and physical carrier sensing indicate the channel is idle. Otherwise, it is in either the back off state or the deference state.

3. AN OPTIMAL FRAME SIZE PREDICTOR

3.1 A Brief of the Kalman Filter

Considering a general prediction problem with noisy data.

$$X_{k+1} = F[X_k; U_{k+1}; W_{k+1}] \tag{1}$$

In this equation, X_k is the system state at time k, U_{k+1} is the system control input at time $k+1$ and W_{k+1} is processing noise at time $k+1$ which is assumed to be additive process noise. $F[\ ]$ is the process model. For the observation system, we have the following equation.

$$Z_{k+1} = H[X_{k+1}; U_{k+1}; V_{k+1}] \tag{2}$$

Here, $H[\ ]$ is the observation model and V_{k+1} is assumed to be additive observation noise. V and W could be any kind of distribution, but generally they are uncorrelated at all time k. In this chapter, our concern is to predict the system states X by using noisy observations Z under the known process model and observation models. When $F[\ ]$ and $H[\ ]$ are linear systems, the Kalman Filter can be used to provide prediction with a least-mean-squared error of true system states recursively.

Kalman filter has been widely utilized for different scenarios like prediction, estimation and smooth. The advantage of Kalman filter is that it is an efficient computational recursive solution of least mean squared error [10, 11]. Kalman filter uses a predictor-corrector structure. The predictor predicts the system state at the next time slot through processing model and the corrector will update the Kalman gain, and then observe the new measurement from the observation model. A posteriori prediction of system state can be derived from Kalman gain, a priori state and the measurement of updated system state. Kalman filter can be represented by the following set of equations.

The processing model is

$$X_{k+1} = AX_k + BU_{k+1} + W_{k+1} \tag{3}$$

and the observation model is

$$Z_{k+1} = HX_{k+1} + V_{k+1} \tag{4}$$

The Kalman gain is

$$K_{k+1} = P_{k+1}H_{k+1}^T(H_{k+1}P_{k+1}H_{k+1}^T + R_{k+1})^{-1} \tag{5}$$

Here, P_{k+1} is a priori prediction error covariance, which can be written as

$$P_{k+1} = E[(X_{k+1} - \bar{X}_{k+1})(X_{k+1} - \bar{X}_{k+1})^T] \tag{6}$$

And the a posteriori update of X_{k+1} is

$$\bar{X}_{k+1} = X_{k+1} + K_{k+1}(Z_{k+1} - H_{k+1}X_{k+1}) \tag{7}$$

3.2 Optimal Frame Size Prediction in the IEEE 802.11 Wireless LAN

The good throughput of the IEEE 802.11 wireless LAN can be expressed by the following equation.

$$T = \frac{L}{(L+H+B+D)(1-P_b)^{L+H} + A} \tag{8}$$

In this equation, L is the payload size of a frame and H is the overhead of transmitting a frame including headers of MAC and PHY and processing delays caused by hardware and software. P_b is the bit error rate under a certain channel quality. B is the average number of time slots of the random back off according to the IEEE 802.11 wireless LAN standard. D and A are respectively the DCF inter-frame space and the equivalent time of an acknowledgement.

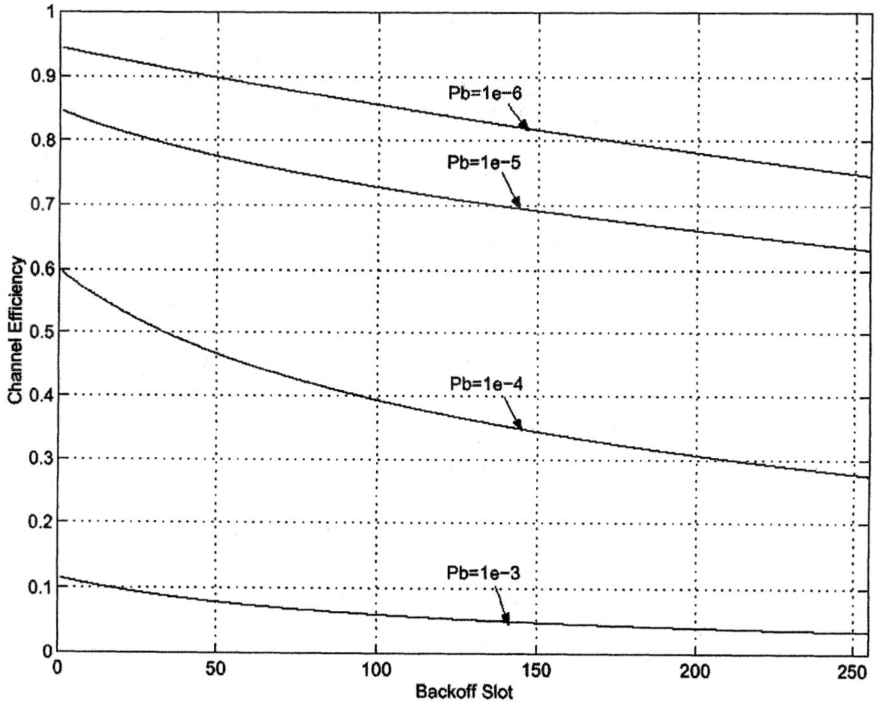

Figure 1. Channel Utilization Vs. Average Backoff Slots under Different Channel Conditions

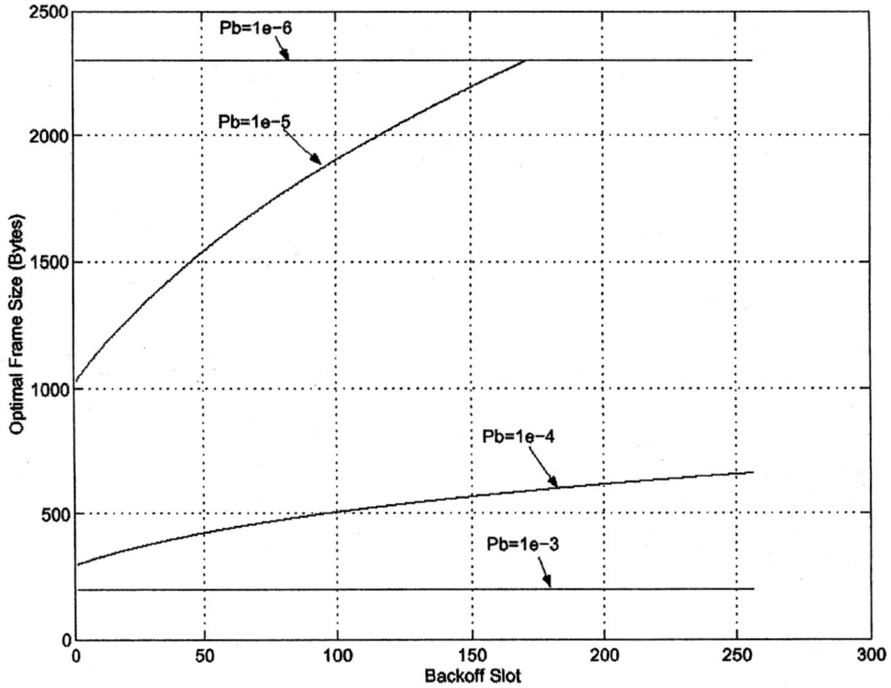

Figure 2. Optimal Payload Size Vs. Average Backoff Slots under Different Channel Conditions

Figure 1 and Figure 2 show relations among bit error rate, frame length and back off time.

Our goal is to maximize the good throughput under a given channel quality. This problem is actually a limitation, a nonlinear optimization problem. Thus, we compute the optimal frame size by maximizing the channel efficiency.

$$\frac{\partial T(L;Pb;B)}{\partial L} = 0 \rightarrow L_{opt} \tag{9}$$

In general, most communication systems have restriction of frame size, L_{max} and L_{min}, thus, the optimal frame size predictor is described below.

The optimal frame size can be computed through taking the derivative of the throughput equation and solve the derived equation. In our case, we can get a close-form solution, which means that it can be easily implemented.

$$L_{opt} = \begin{cases} L_{\max} \leftarrow L_{opt} > L_{\max} \\ L_{opt} \leftarrow L_{\min} < L_{opt} < L_{\max} \\ L_{\min} \leftarrow L_{opt} < L_{\min} \end{cases} \tag{10}$$

The L_{opt} is an optimal frame size but it will be changing since the channel environment is changing in terms of bit error rate, back off time and other factors. Obviously, this predictor is sub-optimal, especially in the case that the channel quality is changing, even though the prediction error will be compensated by the constrained conditions on the frame size.

In order to get more accurate optimal channel prediction, we use Kalman filter to get a more accurate prediction of the optimal frame size used in the next transmission. Thus, following equations are developed as state transition model of the frame size predictor.

$$L_{opt}(k+1) = \begin{cases} L_{\max} \leftarrow L_{opt}(k+1) > L_{\max} \\ L_{opt}(k) + \theta \leftarrow L_{\min} < L_{opt}(k+1) < L_{\max} \\ L_{\min} \leftarrow L_{opt}(k+1) < L_{\min} \end{cases} \tag{11}$$

where, θ is the difference between two optimal frame sizes due to changes of P_b and B. $L_{opt(max)}$ and $L_{opt(min)}$ are respectively maximum and minimum frame size specified by wireless LAN standard. For the observation model, we choose

$$Z(k+1) = L_{opt}(k+1) \tag{12}$$

Z_{k+1} is the observation at time $k + 1$ which will be quantized for efficient process in the system.

4. SIMULATION AND RESULT ANALYSIS

Simulations have been designed to analyze the proposed prediction algorithm and the results are analyzed in this section. The simulation parameters are chosen according to IEEE 802.11b wireless LAN specification (1999). Four modulation schemes are supported in 802.11b physical layer specification, 1Mbps, 2Mbps, 5.5Mbps and 11 Mbps. In this work, without losing generality, the system data rate is set to 11 Mbps with non-FEC CCK modulation scheme. The wireless channel was modeled by a field measurement in a typical office environment [13]. Figure 3 shows the measurements of a typical indoor wireless environment.

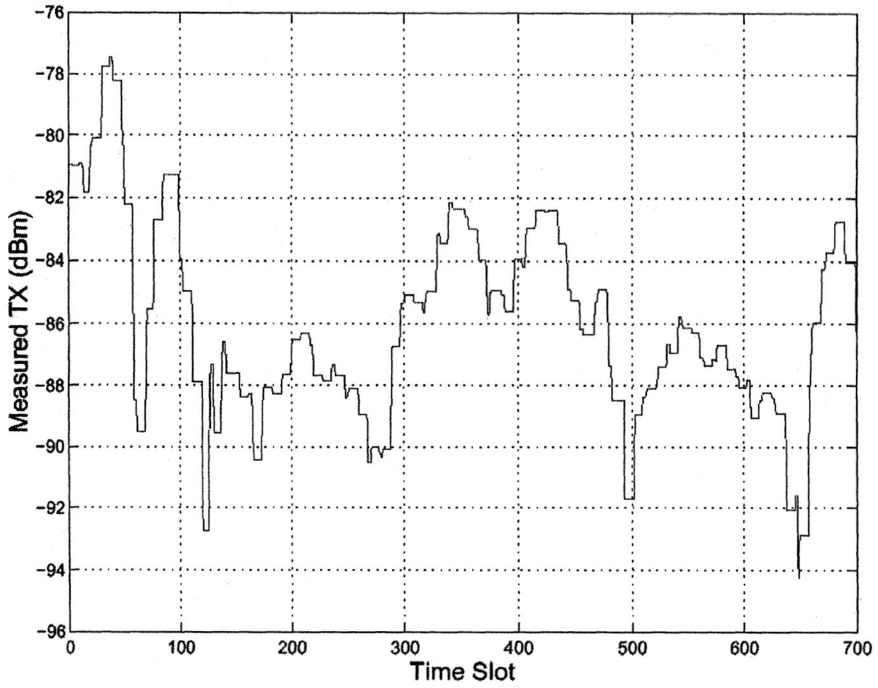

Figure 3. Channel Quality Measurements for a Typical Indoor Environment

L_{max} is 18400 bits; L_{min} is 1150 bits. The average back off time is difficult to calculate due to the memory effect of the random back off algorithm used by CSMA/CA. According to [1,12], each value generated by the back off algorithm of CSMA/CA can be approximately represented by a geometric distribution which takes $p=1/E[B]+1$.

Figure 4 shows the difference between state values and observation values. Another important issue left is to determine values of processing noise and observation noise. Sine here our goal is to verify and analyze the proposed predictor, we use the off-line method to determine these two values. The processing noise is mainly resulted in the equation approximation and the observation noise is mainly from quantization noise. Through off-line observation, the processing error covariance and measure error covariance used in this simulation are taken from the larger variance, which are set to 10000 in this work. For future operational tunings, more observations and measurements could be pursued to determine these two values.

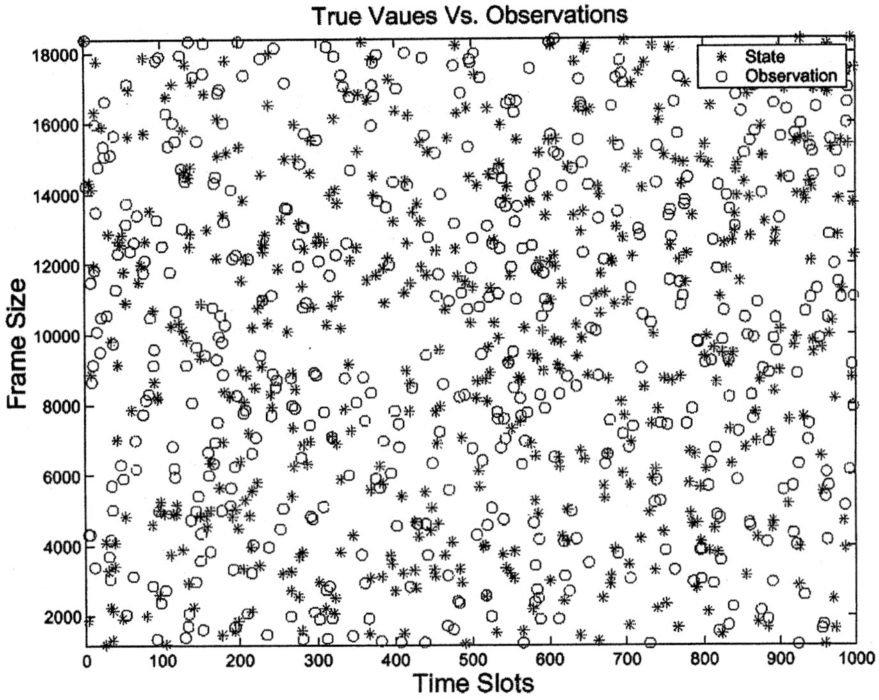

Figure 4. States Vs. Observations

In this work, the simulation runs 10^5 times and the results are compared with results by using moving average method in the Figure 5. In this figure, we make the comparison of different prediction methods. The Root-Mean-Squared (RMS) error of proposed predictor is 588.6 and the RMS error of the moving average method with window size 10 is 12734.7. The root mean squared error of prediction samples is improved by the order of tens with the cost of longer processing time. However, through implementing the proposed predictor in either hardware or firmware, the execution time will be shortened greatly.

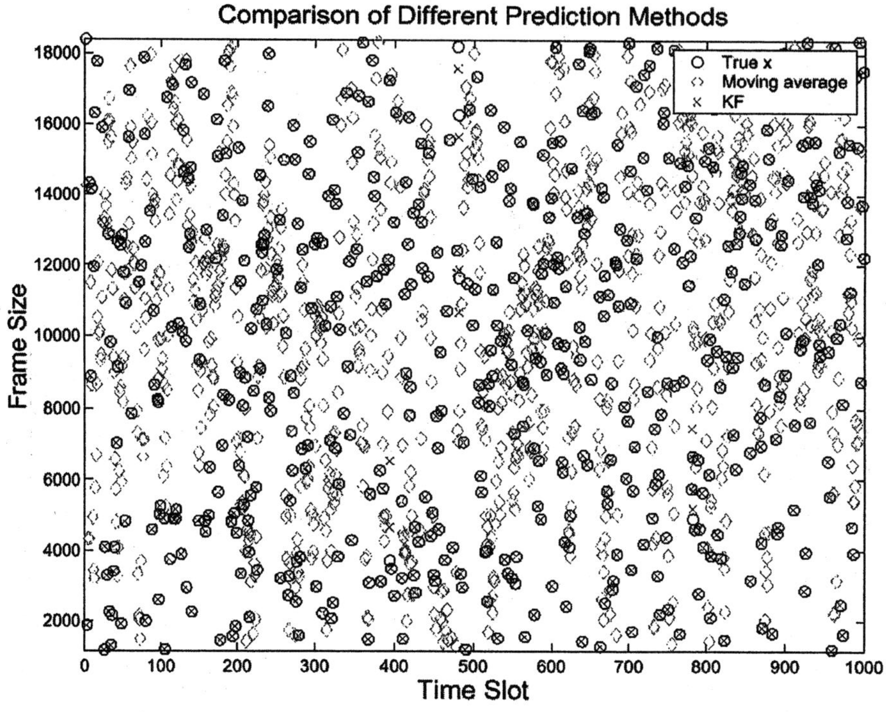

Figure 5. Proposed Predictor Vs. Moving Average with Window Size 10

5. CONCLUSION

In this chapter, we developed an optimal frame size predictor based on Kalman filter. The performance of the proposed predictor was simulated and compared with the performance of the moving average method. The simulation results show that the performance of the new predictor is much better than that of moving average algorithms. This can be concluded from the fact that the RMS error of the proposed predictor is much lower, in the order of hundreds, than that of the moving average method. The processing time is also lowered greatly when it is implemented in firmware or hardware, even though the current processing time is higher than that of the moving average method

REFERENCES

[1] F. Cali, M. Conti, and E. Gregori, "IEEE 802.11 Wireless Lan: Capacity analysis and protocol enhancement," *Proc. IEEE INFOCOM'98, San Francisco, CA, USA*, pp. 142–149, 1998.

[2] C. Chien, M. Srivastava, R. Jain, P. Lettieri, V. Aggarwal, and R. Sternowski, "Adaptive Radio for Multimedia Wireless Links," *IEEE Journal on Selected Areas in Communications*, 17(5), pp. 793–813, 1999.

[3] S. Ci and H. Sharif. Adaptive Approaches to Enhance Throughput of IEEE 802.11Wireless LAN with Bursty Channel," *Proc. of The 25th Annual IEEE Conference on Local Computer Networks (LCN 2000)*, Nov., 2000.

[4] S. Ci, H. Sharif, and A. Young, "Frame Size Prediction for Indoor Wireless Network," *IEE Electronics Letters*, 2001.

[5] A. Goldsmith and S. Chua, "Variable-rate Variable-power MQAM for Fading Channels," *IEEE Transactions on Communications*, 45(10), pp. 1218–30, 1997.

[6] A. Goldsmith and C. S.G, "Adaptive Coded Modulation for Fading Channels," *IEEE Transactions on Communications*, 46(5), pp. 595–602, 1998.

[7] Lettieri P., Fragouli C., and Srivastava M.B, "Low Power Error Control for Wireless Links," *MobiCom '97. Proceedings of the Third Annual ACM/IEEE International Conference on Mobile Computing and Networking*, pp. 139–50, 1997.

[8] E. Modiano, "An Adaptive Algorithm for Optimizing the Packet Size Used in Wireless ARQ Protocols," *Wireless Networks*, 5, pp. 279–286, 1999.

[9] X. Qiu and K. Chawla, "On the Performance of Adaptive Modulation in Cellular Systems," *IEEE Transactions on Communications*, 47(6), pp. 884–95, 1999.

[10] H. Stark and J.Woods, *Probability, Random Processes and Estimation Theory for Engineers*, Prentice Hall, Inc., 1994.

[11] Welch, G. and Bishop, S, "An Introduction to the Kalman Filter," *Technical Report 95-041, University of North Carolina at Chapel Hill*, 1995.

[12] S. Ci, *Link Adaptation for QoS Provisioning in Wireless Data Network*.,Ph.D. Dissertation, 2002.

[13] N. VanErven, L. Sarsoza, and B. Yarbrough, "Fading Channel vs. Frequency Band, Receiver Antenna Height and Antenna Orientation in Office Environment," *Technical Report of Radio System Engineering Group of 3COM Corporation*, 2001.

Chapter 12

PACKET ERROR RATES OF TERMINATED AND TAILBITING CONVOLUTIONAL CODES

Johan Lassing, Tony Ottosson, and Erik Ström

Communication Systems Group, Dept. of Signals and Systems, Chalmers University of Technology, Sweden

Abstract: When a convolutional code is used to provide forward error correction for packet data transmission, the standard performance measures of convolutional codes, i.e., bit error rate and first-event error rate, become less useful. Instead we are interested in the average probability of block (or packet) error. In this chapter a modified transfer function approach is used to obtain a union bound on the block error rate of terminated and tailbiting convolutional codes. The performance of these block codes obtained from termination and tailbiting is compared to the performance of some standard block codes for various information block lengths and also compared to the sphere packing lower bound for optimal block codes. The conclusion is that block codes derived from convolutional codes form a competitive class of short block length block codes.

Key words: convolutional code, block code, block error probability, packet error probability, union bound, terminated convolutional code, tailbiting convolutional code, sphere packing bound.

1. INTRODUCTION

We will consider the case where a rate $R = 1/n$, constraint length K (shift register memory elements + 1) convolutional encoder is used to encode a packet of r binary digits, $\mathbf{q}$. The resulting encoded data packet will consist of $N = nr$ bits (disregarding the termination effects treated in the next section), due to the redundancy added by the convolutional encoder. For a review on the encoding and decoding of convolutional codes, refer to e.g., [1, ch. 4], [2, ch. 11-13]. The encoded block of data will be transmitted to the receiver, that is allowed to observe the N bits, assumed here to have been disturbed independently in each position (dimension) by a zero-mean Gaussian variable of variance $N_0/2$. The average block (or packet) error probability is the probability $\Pr(\hat{\mathbf{q}} \neq \mathbf{q})$ that the receiver maximum-likelihood (ML) decoder decodes the received block into a message $\hat{\mathbf{q}}$ that was not transmitted. In this paper, a union bound on the block error probability will be derived and evaluated to compare certain short length block codes obtained from convolutional codes to some other common block codes of short block lengths.

2. USING A CONVOLUTIONAL CODE AS A BLOCK CODE

The standard application of convolutional codes is in situations where a low bit error probability (as opposed to block error probability) is desired, despite bad channel conditions. Since no requirement is put on the block error probability, the transmitted sequences may be very long, so that the probability that the entire sequence will be decoded correctly goes to zero. The average bit error probability will however still be small for well-designed codes and reasonable bit-energy-to-noise ratios. In packet transmission the situation is quite different; a small amount of data bits (in the order of 100 bits) are to be transmitted and decoded without errors or the receiver will request a repetition.

Several strategies to turn a convolutional code into a block code can be distinguished between among which termination and tailbiting are two practical methods [3]. When zero-tail *termination* is applied the convolutional encoder is initially in a known state, which we assume to be the zero state, and the message sequence is padded at the end by $K-1$ zeros, so that the encoder returns to the zero state after having encoded the message. The price that is paid for this is a fractional rate loss, since $K-1$ known information bits are added to the original message. If the original information block contained r bits, the output codewords will be of length $(r+K-1)n$, so that the effective rate becomes

$$R' = R\left(1 - \frac{K-1}{r+K-1}\right) \tag{1}$$

where the term $K-1/r+K-1$ is the fractional rate loss and goes to zero as $r \to \infty$.

The method of *tailbiting* [4] avoids the problem of the fractional rate loss by letting the last $K-1$ information bits define the starting state of the encoder and then clocking the encoder r times, so that exactly rn encoded bits are produced. This means that no rate loss occurs and that the encoding always starts and ends in the same state, but not necessarily the zero state. The price that is paid is the increased decoder complexity needed to estimate the initial state of the encoder and a poorer weight distribution of the derived block code (see section 4).

If decoding complexity is not our main concern, we may also consider the case where we let the last $K-1$ bits define the starting state (as for tailbiting), but we only clock the encoder $r-K$ times. This means that the encoder can start and end in any state, as opposed to the above mentioned methods. This will increase the decoding complexity, and also increase the probability of decoder error. However, we now obtain what we may call a fractional rate *gain*, which will counteract the increased error probability. The effective for this method becomes

$$R' = R\left(1+\frac{K}{r-K}\right) \tag{2}$$

and since now $R' > R$, this will counteract the increased probability of error.

All the above outlined termination methods result in a number of valid code sequences or codewords. By listing all these codewords we obtain a set C, which is called the codebook. This means that the convolutional code is turned into a block code.

Only zero-tail termination and tailbiting will be considered in this paper, while results for the third method are not included. In section 3, the union bound on the error probability of a block code is revised and a method of obtaining the necessary weight distributions is suggested in section 4. In the final sections an application to packet data systems is evaluated and results and comments are given.

3. UNION BOUND ON THE ERROR PROBABILITY OF BLOCK CODES

A common way to estimate the performance of a block code is to use the union of the pair-wise error probabilities of all codewords in the block code. For the case of a white, Gaussian disturbance in each dimension, the pair-wise error probability P_2 of two codewords $\mathbf{a}$ and $\mathbf{b}$, with Hamming distance $d_H(\mathbf{a},\mathbf{b}) > 0$, is given as (assuming binary or quaternary phase-shift modulation)

$$P_2 = \Pr\{\mathbf{a} \neq \mathbf{b}\} = Q\left(\sqrt{2d_H(\mathbf{a},\mathbf{b})\gamma_b/n}\right) \tag{3}$$

where the Q-function is given by

$$Q(x) = \frac{1}{\sqrt{2\pi}} \int_x^{\infty} \mathrm{e}^{-t^2/2} dt \tag{4}$$

The probability P_2 is the probability that the ML receiver decides in favor of codeword $\mathbf{b}$ when codeword $\mathbf{a}$ was transmitted and only those two codewords exist in the communication system codebook. In (3), $\gamma_b = E_b/N_0$ is the bit-energy-to-noise ratio of the bits entering the decoder, where E_b is the bit energy and N_0 is the one-sided power spectral density of the white noise.

The average block-error probability, P_e, of the ML receiver is given by

$$\begin{aligned} P_e &= \sum_{\mathbf{a}\in C} \Pr\{\mathbf{a}\text{ transmitted}\} \Pr\{\mathbf{a}\text{ not decoded}|\mathbf{a}\text{ transmitted}\} \\ &= \sum_{\mathbf{a}\in C} \Pr\{\mathbf{a}\text{ transmitted}\} \sum_{\mathbf{b}\in C,\, \mathbf{b}\neq\mathbf{a}} \Pr\{\hat{\mathbf{b}} = \mathbf{b}|\mathbf{a}\text{ transmitted}\} \end{aligned} \tag{5}$$

where $\hat{\mathbf{b}}$ is the decoder decision on the transmitted *encoded* block. Since, there is a one-to-one (i.e., error-free) mapping between the block $\hat{\mathbf{b}}$ and $\hat{\mathbf{q}}$, the above expression is indeed the average block error probability of ML decoder output. It is common and reasonable to assume that each codeword is equally likely to be transmitted, so that $\Pr\{\mathbf{a}\} = 2^{-r}$. However, the main problem with (5) is the probability terms in the second sum, $\Pr\{\hat{\mathbf{b}} = \mathbf{b}|\mathbf{a}\text{ transmitted}\}$. In most cases of interest, there is no tractable way of obtaining these probabilities, so instead we will try to obtain an upper bound on this sum. One way to do this is to consider the union of the pair-wise error probabilities of the transmitted vector $\mathbf{a}$ to all other codewords. This yields,

$$
\begin{aligned}
P_e &\le \sum_{\mathbf{a}\in\mathrm{C}} \Pr\{\mathbf{a}\text{ transmitted}\} \sum_{\mathbf{b}\in\mathrm{C},\, \mathbf{b}\neq\mathbf{a}} Q\left(\sqrt{2d_H(\mathbf{a},\mathbf{b})\gamma_b/n}\right) \\
&= 2^{-r} \sum_{\mathbf{a}\in\mathrm{C}} \sum_{\mathbf{b}\in\mathrm{C},\, \mathbf{b}\neq\mathbf{a}} Q\left(\sqrt{2d_H(\mathbf{a},\mathbf{b})\gamma_b/n}\right)
\end{aligned} \tag{6}
$$

To come further, we use the fact that our convolutional code is linear, and therefore the derived block code is also linear. A linear block code has the appealing property that the sum (over $GF(2)$) of any two codewords is also a codeword, i.e., $\mathbf{a}+\mathbf{b}=\mathbf{c};\ \mathbf{a},\mathbf{b},\mathbf{c}\in\mathrm{C}$. In particular, the all-zero codeword must be part of any linear code. Linearity gives us that all 2^{-r} terms in the inner sum over the codewords in (6) are equal, so that for linear block codes (6) simplifies to

$$
P_e \le \sum_{\mathbf{b}\in\mathrm{C},\, \mathbf{b}\neq\mathbf{a}} Q\left(\sqrt{2d_H(\mathbf{0},\mathbf{b})\gamma_b/n}\right) \tag{7}
$$

where we have assumed that the all-zero codeword is transmitted. We observe that the only code dependent parameter in this expression is the function $d_H(\mathbf{0},\mathbf{b})$, the Hamming weight of the binary vector $\mathbf{b}$. Therefore, if we define the weight distribution function $w(d)$ as the number of codewords in C having Hamming weight d, we may further simplify the union bound of (7) to

$$
P_e \le \sum_{d=1}^{d_{\max}} w(d)\, Q\left(\sqrt{2d\gamma_b/n}\right) \tag{8}
$$

where $d_{\max}$ is the maximum Hamming weight of any codeword in C. Note that the sum starts at $d=1$ to exclude the transmitted all-zero codeword. For zero-tail terminated convolutional codes, $d_{\max} \le (r+K-1)n$, while for tailbiting convolutional codes, $d_{\max} \le rn$.

The evaluation of (8) only requires knowledge of the weight distribution $w(d)$ of the code, which is significantly easier to obtain than finding closed forms for the expressions we started with, i.e., (5). The price we have paid for this simplification is that (8) is an upper bound on the error probability and therefore, at best, it is a good approximation to the true error probability. In general, it is necessary to do computer simulations to establish over what range this approximation is valid.

4. FINDING THE WEIGHT DISTRIBUTION

The generation of output sequences from a convolutional encoder is conveniently described by the encoder state diagram. In figure 1 the state diagram of an $R=1/2, K=3$, maximum free distance convolutional code [5] is shown.

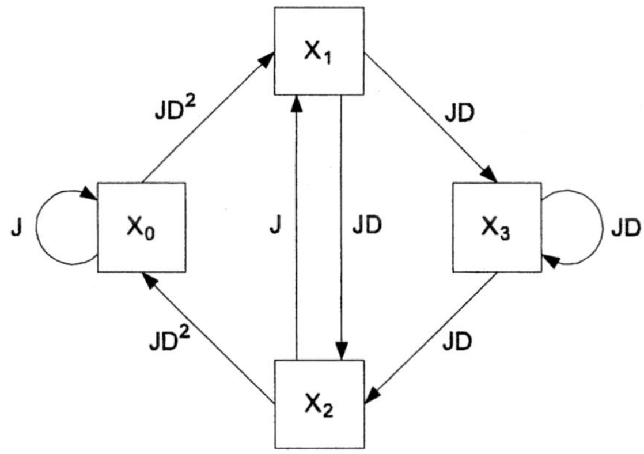

Figure 1. State diagram for a $R=1/2,\ K=3$ convolutional encoder.

The state diagram is seen to be a weighted, directed graph, where the weights on the graph edges (transitions between states) are called indeterminates and represent some property that will accumulate as the vertices (states) are traversed. In the example diagram of figure 1, J enumerates how many transitions that are made and D enumerates the Hamming weight of the encoded bits on each transition. The exponents of D are encoder properties and need to be carefully selected to obtain good codes (usually through computer search, see e.g., [5]).

For the present problem we need to be able to find the path enumerators of all paths through the state diagram starting and ending in the same state X_i after L transitions, where $L=r+K-1$ for terminated convolutional codes and $L=r$ for tailbiting convolutional codes. To find the path enumerators, we modify the standard transfer function method of finding the distance properties of convolutional codes as described in section 4.3 of [1]. In the modified scheme, the enumerators for all paths starting and ending in state X_i are obtained by adding to the state diagram a starting and ending state and connecting them appropriately. This is illustrated for state X_0 in figure 2 for the example encoder. Since the encoding is constrained to start in state X_0, the state δ is introduced. The connections from δ to the other states are the same as for X_0 (refer to figure 1). In the same way, the state X_4 is introduced as a collecting state. By letting the starting state $\delta=1$, the ending state X_4 will account for all paths through the state diagram starting and ending in state X_0. Note that all possible code sequences starting and ending in state X_0 are counted and not only unmerged sequences, as is normal for evaluating convolutional code performance. To further clarify the procedure, the modified state diagram used to find the enumerators for all paths starting and ending in state X_1 is shown in figure 3. Note that the original state diagram remains unchanged and the only thing that changes are the connections (and the weights) from the added starting and ending states.

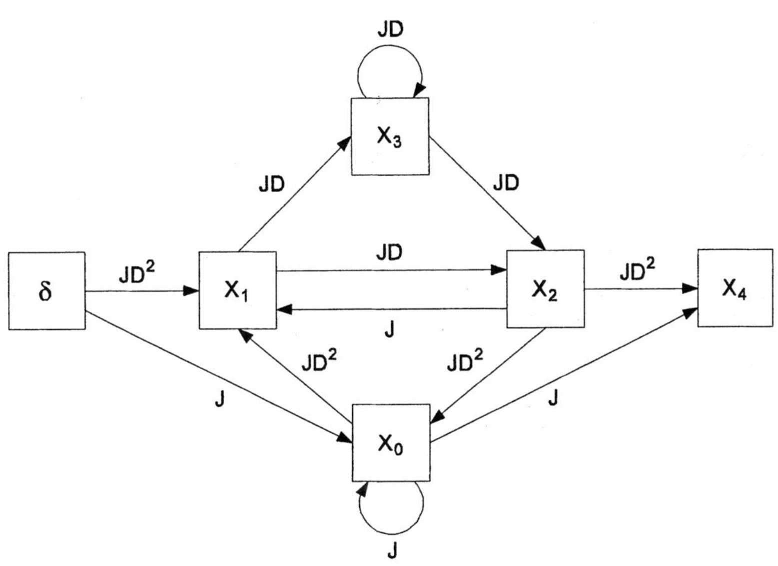

Figure 2. Modified state diagram for a $R = 1/2, K = 3$ convolutional encoder. X_4 enumerates all paths starting and ending in state X_0.

The modified state diagrams in figures 2 and 3 each describes a system of equations in $5\left(2^{K-1}+1\right)$ unknowns (one for each node, minus the initial state, since $\delta = 1$),

$$\mathbf{M}_i \mathbf{x}_i = \mathbf{r}_i \tag{9}$$

where $\mathbf{M}_i$ is an $\left(2^{K-1}+1\right)\times\left(2^{K-1}+1\right)$ matrix and the index $i \in \left\{0,1,\ldots,2^{K-1}-1\right\}$ indicates for what starting and ending state X_i the solution is valid. The right-hand side vector, $\mathbf{r}_i$, is a vector containing the weights on the transitions from the starting state δ.

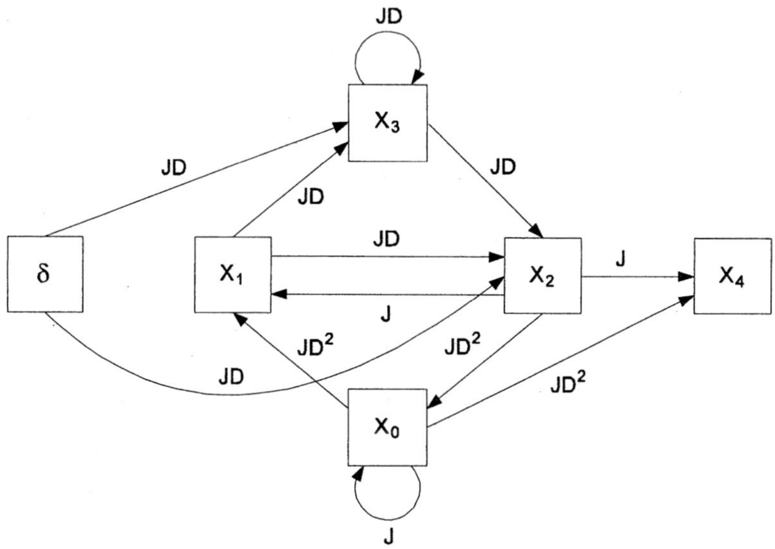

Figure 3. Modified state diagram for a $R = 1/2, K = 3$ convolutional encoder. X_4 enumerates all paths starting and ending in state X_1.

It is illustrative to explain how the matrix equation in (9) is obtained. The starting point is the state transition matrix, i.e., the matrix containing the weights of all transitions of the state diagram in figure 1. For this example encoder, the state transition matrix is

$$\mathbf{A} = \begin{bmatrix} J & JD^2 & 0 & 0 \\ 0 & 0 & JD & JD \\ JD^2 & J & 0 & 0 \\ 0 & 0 & JD & JD \end{bmatrix} \tag{10}$$

where matrix element a_{kl} is the path enumerator for the transition from state k to state l (the path enumerator is set to zero if no such transition is available). Note that this indexing convention is slightly different from what is normal for matrices, since the upper left element with this convention is a_{00} and not a_{11} as is more common. Let the columns of the matrix $\mathbf{A}$ be labeled by $\mathbf{A}[j]$ and the rows by $\mathbf{A}'[j]$, where the first column (or row) is given by $j = 0$. Next, if we are interested in all paths starting and ending in a particular state, $\mathbf{X}_i$, we form a new matrix $\mathbf{B}_i$ by appending to $\mathbf{A}$ the ith column of $\mathbf{A}$ and making it square by appending a row of $2^{K-1}+1$ zeros. This means that this new matrix has the form

$$\mathbf{B}_i = \begin{bmatrix} \mathbf{A} & \mathbf{A}[i] \\ \mathbf{0} & 0 \end{bmatrix} \tag{11}$$

From $\mathbf{B}_i$ the matrix $\mathbf{M}_i$ is obtained as

$$\mathbf{M}_i = \mathbf{I} - \mathbf{B}_i' \tag{12}$$

where $\mathbf{I}$ is the identity matrix and the prime indicates matrix (or vector) transpose. The right hand side vector of (9), $\mathbf{r}_i$, is simply given by

$$\mathbf{r}_i = \begin{bmatrix} \mathbf{A}'[i] & 0 \end{bmatrix}' \tag{13}$$

For zero-tail (zt) terminated convolutional codes we are interested in the solution for $i = 0$, so that for our present example we obtain

$$\begin{bmatrix} 1-J & 0 & -JD^2 & 0 & 0 \\ -JD^2 & 1 & -J & 0 & 0 \\ 0 & -JD & 1 & -JD & 0 \\ 0 & -JD & 0 & 1-JD & 0 \\ -J & 0 & -JD^2 & 0 & 1 \end{bmatrix} \begin{bmatrix} X_0 \\ X_1 \\ X_2 \\ X_3 \\ X_4 \end{bmatrix} = \begin{bmatrix} J \\ JD^2 \\ 0 \\ 0 \\ 0 \end{bmatrix} \tag{14}$$

From the solution of this system of equations, the desired transfer function is obtained from $X_4[i=0] \equiv X_4[0]$, where we have emphasized that the solution was obtained for paths starting and ending in state X_0. In this case,

$$T_{\text{zt}}(D,J) = X_4[0] = \frac{J^2 - (J^3 + J^4)D + (J^3 + J^4)D^5}{1 - J - (J - J^3)D - J^3 D^5} \tag{15}$$

Since the indeterminate J enumerates the number of transitions, the weight spectrum of a block code obtained from zero-tail termination after L transitions is found from the coefficient of the factor J^L of the serial expansion of $T_{\text{zt}}(D,J)$. This coefficient, which is a polynomial in D, is the weight distribution of the block code. For example, a block code obtained from $L = 6$ transitions (information block length $r = L - K + 1 = 4$) in the modified state diagram of figure 2 has weight distribution polynomial

$$W_6(D) = 1 + 4D^5 + 5D^6 + 4D^7 + D^8 + D^{10} \tag{16}$$

For tailbiting codes, we are interested in the weights of the *collection* of all paths starting and ending in each of the states X_i after L transitions. There are several possible ways of obtaining this collection, but the straightforward approach is to realize that, since all paths starting and ending in X_i are accounted for by the last element ($X_4[i]$) of the solution vector x_i, the collection of paths is accounted for by

$$T_{\text{tb}}(D,J) = \sum_{i=0}^{2^{K-1}-1} X_4[i] \tag{17}$$

so that the system of equations in (9) needs to be solved 2^{K-1} times to get $T_{\text{tb}}(D,J)$. The weight distribution of the block code formed from the tailbiting convolutional code is now found in exactly the same way as for the zero-tail termination case, as described above. For example, the weight distribution of a block code obtained from tailbiting of the example convolutional code discussed above with $L = 4$ transitions (information block length $r = L = 4$) is given in the third column of table 1. It is seen from the table that the free distance of the tailbiting code is only 2, whereas it is 5 for the terminated code of the same information block length, but this apparent drawback is counteracted by the fractional rate loss of the terminated code, which reduces the effective bit-energy-to-noise ratio for the terminated code.

Other ways of obtaining the weight distributions discussed above and related weight distributions can be found in [3, 4].

5. APPLICATION TO PACKET DATA SYSTEMS

As an interesting application of the methods suggested above, let us consider a communication system where data is transmitted in small bursts of less than 100 information bits at a time. We assume that each data packet is encoded by a rate 1/2 encoder (block or convolutional), but that no additional latency is tolerated. This means that the encoder is not allowed to wait for additional packets to arrive and treat these as longer blocks during encoding.

Apart from the terminated and tailbiting convolutional codes, we also assume three different quadratic residue (QR) block codes to be used, all of rate 1/2; the Hamming (8,4,4), the Golay (24,12,8) and the quadratic residue (48,24,12) codes [6, ch. 16]. For the block codes, the notation (N, m, d) is used, where N is the coded block length, m is the input information block length and d is the smallest Hamming distance between any two codewords in the code. For convolutional codes, the notation is (n, k, K), where n and k are the number of input bits and coded output bits per cycle, respectively, and K, as before, is the constraint length.

The block error rates of the QR codes will be compared to the block error rates of terminated and tailbiting convolutional codes of various constraint lengths using the methods described in section 4. For all codes, maximum-likelihood decoding is assumed, where the complexity (in terms of trellis edge evaluations per decoded bit) of ML decoding of the QR block codes above is roughly comparable to ML decoding of convolutional codes with constraint lengths 3, 7 and 15, respectively [7, 8].

6. RESULTS AND SIMULATIONS

We start out by giving the detailed calculations for the simplest codes studied here; the Hamming (8, 4, 4) code and the terminated and tailbiting convolutional codes of constraint length $K = 3$. In Table 1 the weight distribution of the Hamming (8, 4, 4) code [6, p. 597] is given along with the weight distributions for the Golay (24, 12, 8) [6, p. 597] and the quadratic residue (48, 24, 12) [6, p. 604] codes.

Table 1. Weight distributions of Hamming (8,4,4), Golay (24,12,8) and quadratic residue (48,24,12) block codes.

d	$w_H(d)$	$w_G(d)$	$w_{QR}(d)$
0	1	1	1
4	14	0	0
8	1	759	0
12	0	2576	17296
16	0	759	535095
20	0	0	3995376
24	0	1	7681680
28	0	0	3995376
32	0	0	535095
36	0	0	17296
48	0	0	1

If we assume that the information packet consists of $r = mp$ bits, where $p \in \mathbf{Z}^+$ and $m = 4$ for the Hamming (8, 4, 4) code, the encoder will output p blocks of $N = 8$ encoded bits each. For the entire encoded packet of $8p$

bits to be received without error, each of the p encoded blocks must be correctly received. Therefore, the probability that the packet is received in error is given by

$$P_w = 1 - (1 - P_e)^p \tag{18}$$

where P_e is the probability that an encoded sub-packet is erroneously decoded, which is given by (8) and evaluated using the weight distribution in the second column of Table 2.

For a terminated convolutional code, on the other hand, the entire information block of r bits can be encoded and only in the end it needs to be terminated by appending $K-1$ zeros. The effective rate of a terminated rate 1/2 convolutional code of constraint length K as a function of r is according to

$$R' = \frac{r}{2(r+K-1)} = \frac{1}{2} - \frac{K-1}{2(r+K-1)} \tag{19}$$

so that $R' \to 1/2$ as $r \to \infty$, as expected. As r gets larger, the number of codewords will of course increase, but the free distance of the block code derived from termination will not change. Instead, the multiplicities of the low order terms of $w(d)$ increase, as can be seen from Table 2. In this table, the weight distributions for $r=4$, $r=6$ and $r=8$ information bits per encoded block is given. For the block codes derived from tailbiting the free distances start at low values, increasing until the free distance is the same as for the terminated codes. To make fair comparisons between the block codes, the tailbiting convolutional codes and the terminated convolutional codes, the effective bit-energy will be lowered for the terminated convolutional codes according to the rate reduction of (1).

In figure 4 the bit-energy-to-noise ratio required to achieve $P_w = 10^{-4}$ (or $P_e = 10^{-4}$ for convolutional codes) on an additive, white, Gaussian noise (AWGN) channel is plotted as a function of the information block size for the QR codes and the $K=3$ tailbiting and terminated convolutional codes. The performance of the QR codes is evaluated using (18) and for the convolutional codes (8) is used. Also included for comparison is the sphere packing lower bound (spb) on the performance of the optimal codes for each r [7, 9, 10, 11]. This bound is a better alternative for comparison in this case (and many others) than the channel capacity (which gives $E_b / N_0 = 0$ dB for this case), since the channel capacity is assuming infinite block length.

The bound shown is valid for the continuous-input AWGN channel, which makes it slightly too optimistic in the comparison made here.

Table 2. Weight distributions of block codes obtained via termination (zt) and tailbiting (tb) of a $R=1/2, K=3$ convolutional code for $r=4,6$ and 8 information bits.

d	$w(d)$					
	$r=4$		$r=6$		$r=8$	
	zt	tb	zt	tb	zt	tb
0	1	1	1	1	1	1
1	0	0	0	0	0	0
2	0	2	0	0	0	0
3	0	4	0	2	0	0
4	0	1	0	9	0	2
5	4	4	6	12	8	24
6	5	4	9	13	13	36
7	4	0	12	18	20	40
8	1	0	12	6	28	57
9	0	0	6	0	32	40
10	1	0	3	3	38	20
11	0	0	8	0	40	24
12	0	0	3	0	40	12
13	0	0	0	0	24	0
14	0	0	0	0	5	0
15	0	0	0	0	4	0
16	0	0	0	0	3	0
R'	$1/3$	$1/2$	$3/8$	$1/2$	$2/5$	$1/2$

It is obvious from figure 4 that the QR codes all perform very well for their shortest block lengths (within 0.7 dB of an optimal code), but they all start to deviate as the block length is increased. The terminated and tailbiting convolutional codes also perform well at short block lengths, the terminated code being slightly superior for the very shortest block lengths, but as the block lengths increase, the performance curves flattens out. It is interesting to note that the slope of all the curves as the block length increases is very similar.

Regarding the decoding complexity of the codes, the decoding complexity of the Hamming (8,4,4) code and the terminated code is similar, whereas the ML decoding of the tailbiting code requires one pass of the Viterbi algorithm for each state (2^{K-1}), so that it is 4 times more complex for this case. However, at least for block lengths above 50, the gain of the tailbiting code is marginal over the terminated code, and the increased complexity is perhaps not motivated. Note that there are several sub-optimal decoding procedures suggested for tailbiting codes (see e.g., [4]), that would reduce the complexity difference (but also the coding gain), but this is not considered here.

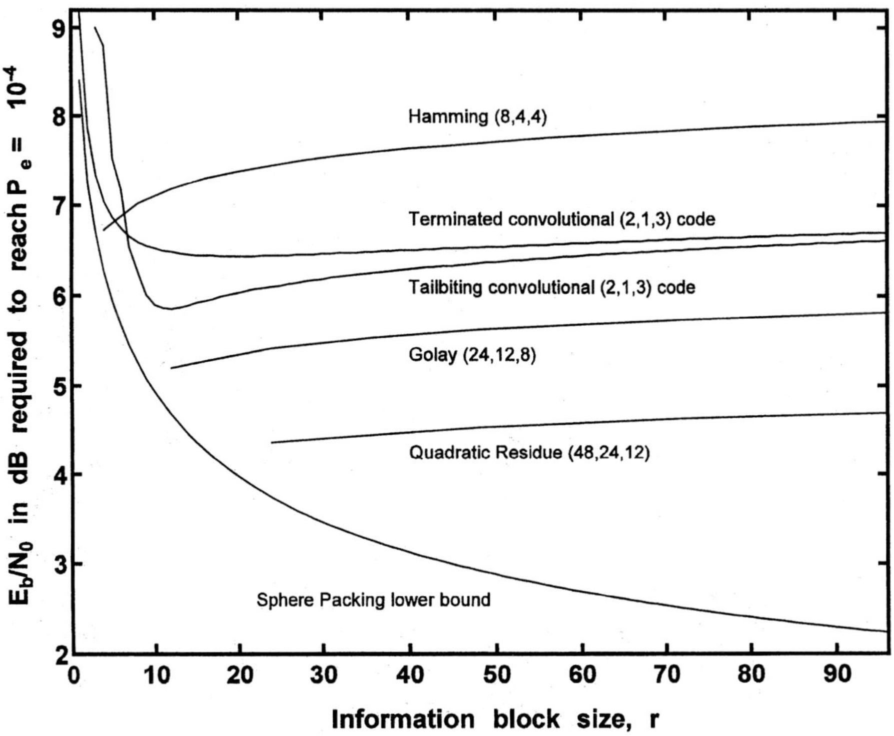

Figure 4. Plot showing the bit-energy-to-noise ratio required to achieve $P_W = 10^{-4}$ on an AWGN channel for various short $R=1/2$ block codes. The terminated and tailbiting convolutional codes have constraint length $K=3$. The information block size ranges from 1 to 96. Included for comparison is the sphere packing lower bound, which is a lower bound on the performance of the optimal codes.

In figure 5 the same comparison is made, but the constraint length of the convolutional code is increased to $K = 5$ and the interesting region is zoomed in on, so that r ranges from 1 to 48. The decoding complexity of this convolutional code is much less than the Golay (24, 12, 8) code, but still it is seen that the tailbiting code outperforms the Golay code over almost the entire range of block lengths. For block lengths above 40 both the terminated and the tailbiting codes perform better. Another interesting observation is made in this figure, the heavy variations of the required E_b / N_0 for the tailbiting codes of shorter block lengths. This is a result of the truncation effects that occur when the paths through the state diagram are required to start and end in the same state (and also an effect of the fact that to obtain the best results we shall use a different set of code generators for different r).

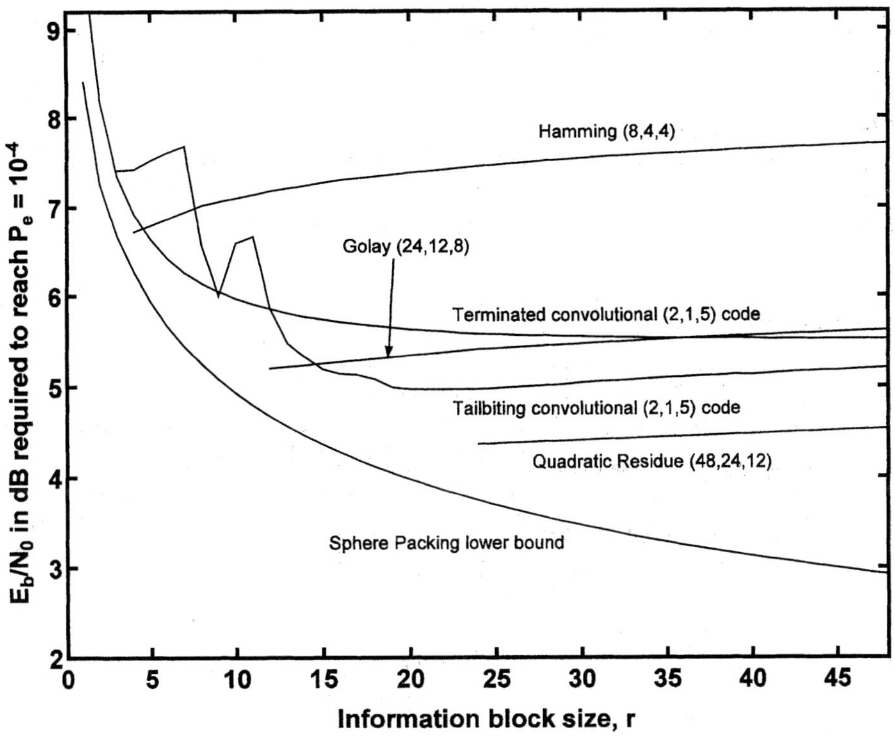

Figure 5. Plot showing the bit-energy-to-noise ratio required to achieve $P_W = 10^{-4}$ on an AWGN channel for various short $R = 1/2$ block codes. The terminated and tailbiting convolutional codes have constraint length $K = 5$. The information block size ranges from 1 to 48. Included for comparison is the sphere packing lower bound, which is a lower bound on the performance of the optimal codes.

7. CONCLUSIONS

A method to calculate the union bound on the average block-error probability of a block code obtained from a zero-tail terminated or tailbiting convolutional code is given. The only code dependent quantity needed to evaluate the performance is the weight distribution of the derived block code, and a method to obtain the weight distribution of both terminated and tailbiting convolutional codes is given. The proposed method involves a minor modification to the well known procedure to obtain the transfer function of a convolutional code given in, e.g., [1]. The union bound was applied to terminated and tailbiting rate 1/2 convolutional codes of constraint lengths $K = 3$ and $K = 5$ and the performance was compared to that of three common rate 1/2 quadratic residue codes. For comparable complexity, it was seen that the block codes derived from convolutional codes outperformed the QR codes and that the tailbiting codes are superior to the terminated codes, except for the very shortest block lengths. However, when the information block lengths exceed 50 bits, the energy gain becomes marginal. It is concluded that block codes derived from convolutional codes forms an attractive class of block codes that perform very well compared to the QR block codes of similar decoding complexity studied in this chapter.

REFERENCES

[1] A. Viterbi and J. K. Omura, *Principles of digital communication and coding*, McGraw-Hill, New-York, 1979.

[2] S. Wicker, *Error control systems for digital communication and storage*, Prentice-Hall, Upper Saddle River, NJ, 1995.

[3] J. K. Wolf and A. J. Viterbi, "On the weight distribution of linear block codes formed from convolutional codes," *IEEE Transactions on Communications*, vol. COM-44, no. 9, pp. 1049–1051, Sept. 1996.

[4] H. H. Ma and J. K. Wolf, "On tail biting convolutional codes," *IEEE Transactions on Communications*, vol. COM-34, no. 2, pp. 104–111, Feb. 1986.

[5] P. Frenger, P. Orten, and T. Ottosson, "Convolutional codes with optimum distance spectrum," *IEEE Communications Letters*, vol. 3, no. 11, pp. 317–319, Nov. 1999.

[6] F. J. MacWilliams and N. J. A. Sloane, The *theory of error-correcting codes*, North-Holland, Amsterdam, 1977.

[7] S. Dolinar, D. Divsalar, and F. Pollara, "Code performance as a function of block size," *JPL TDA Progress Report 42-133*, May 1998.

[8] A. Kiely, S. Dolinar, R. McEliece, L. Ekroot, and W. Lin, "Trellis decoding complexity of linear block codes," *IEEE Transactions on Information Theory*, vol. 42, no. 6, pp. 1687–1697, Nov. 1996.

[9] C. E. Shannon, "Probability of error for optimal codes in a Gaussian channel," *Bell System Technical Journal*, vol. 38, pp. 611–656, May 1959.

[10] D. Slepian, "Bounds on communication," *Bell System Technical Journal*, vol. 42, pp. 681–707, May 1963.

[11] S. J. MacMullan and O. M. Collins, "A comparison of known codes, random codes, and the best codes," *IEEE Transactions on Information Theory*, vol. 44, no. 7, pp. 3009–3022, Nov. 1998.

Chapter 13

THE FENG-RAO DESIGNED MINIMUM DISTANCE OF BINARY LINEAR CODES AND CYCLIC CODES

Junru Zheng*, Takayasu Kaida**, Kyoki Imamura*
** Kyushu Institute of Technology, ** Yatsushiro National College of Technology*

Abstract: The definition of the Feng-Rao designed minimum distance, first introduced into algebraic geometry codes, has been extended to the case of general linear codes by Miura. In Miura's definition the Feng-Rao designed minimum distance d_{FR} is determined by an ordered basis related to a given linear code over a finite field F_q. Matsumoto gave a generalized definition d^*_{FR} of d_{FR} with three ordered bases. We have Miura's definition if three ordered bases are same in Matsumoto's definition. In this Chapter, firstly, it is shown that the Feng-Rao designed minimum distance of binary linear codes cannot take an odd value except one if we use Miura's definition. Secondly, from some properties and some numerical examples of Matsumoto's d^*_{FR} we conjecture that Matsumoto's generalization is not so effective for binary linear codes compared with Miura's definition. Thirdly, the Type I ordered basis is introduced for computing d_{FR} of cyclic codes. It is shown that the Type I ordered basis is worst choice in case of nonbinary cyclic codes, although it is not so bad in case of binary $(n, 1)$, $(n, 2)$ and $(n, n\text{-}1)$ cyclic codes.

Key words: binary linear code, Feng-Rao designed minimum distance, ordered basis, Miura's definition, Matsumoto's definition

1. INTRODUCTION

The Feng-Rao designed minimum distance and the Feng-Rao decoding were originally introduced into algebraic geometry codes [1]. They have been extended to the case of general linear codes over a finite field by Miura [3]. Miura's definition of the Feng-Rao designed minimum distance d_{FR} for an (n, k) linear code C over a finite field F_q of order q depends on the choice of an ordered basis $W_n = \{\boldsymbol{w}_1, \boldsymbol{w}_2, \cdots, \boldsymbol{w}_n\}$ of the vector space F_q^n with dimension n over F_q. It is interesting to find such an optimum ordered basis W_n as d_{FR} is maximum, since the Feng-Rao decoding algorithm can correct up to $[(d_{FR}-1)/2]$ errors. Recently the authors showed some properties of the Feng-Rao designed minimum distance d_{FR} by Miura for binary linear codes and cyclic codes [4, 5].

Recently the definition of d_{FR} of linear codes has been slightly generalized by Matsumoto and Miura [2], which uses three ordered bases $U_n = \{\boldsymbol{u}_1, \boldsymbol{u}_2, \cdots, \boldsymbol{u}_n\}$, $V_n = \{\boldsymbol{v}_1, \boldsymbol{v}_2, \cdots, \boldsymbol{v}_n\}$ and $W_n = \{\boldsymbol{w}_1, \boldsymbol{w}_2, \cdots, \boldsymbol{w}_n\}$ of F_q^n instead of one in case of Miura's definition, i.e., W_n is used for defining the linear code, and U_n, V_n are used for computing a syndrome matrix in Matsumoto's definition and Miura's definition is a special case of $U_n = V_n = W_n$.

In this Chapter, firstly we show the Feng-Rao designed minimum distance of binary linear codes can not take an odd value except one if we use Miura's definition.

Secondly, we discuss the choice of three ordered bases by Matsumoto's definition for binary linear codes. Consequently we conjecture that Matsumoto's generalization for binary linear codes is not so effective compared with Miura's definition from some properties and some numerical examples of Matsumoto's d_{FR}.

Thirdly, the Type I ordered basis W_n was introduced as a very natural candidate necessary for computing d_{FR}. The ordered basis W_n is Type I if its subset W consists n-k row vectors of the permutation of the usual parity check matrix defined by parity check polynomial of a cyclic code. It was shown that the choice of an ordered basis with Type I is worst in many cases of nonbinary cyclic codes, since the Feng-Rao designed minimum distance is equal to 1 if the check polynomial has a coefficient neither equal to 0 nor 1. It is also shown that in the case of binary (n, k) cyclic codes C with $k = 1, 2$ and n-1 the choice of an ordered basis with Type I is not so bad, since there exists an ordered basis with Type I such that the Feng-Rao designed minimum distance is equal to n-1, $2(n/3-1)$, and 2, respectively.

2. DEFINITIONS OF THE FENG-RAO DESIGNED MINIMUM DISTANCE

In this section we will briefly review results of [1] and [2] necessary for the following discussions.

For two vectors $\boldsymbol{x} = \{x_1, x_2, \cdots, x_n\}$ and $\boldsymbol{y} = \{y_1, y_2, \cdots, y_n\} \in F_q^n$, their product $\boldsymbol{xy}$ is defined as $\boldsymbol{xy} = \{x_1y_1, x_2y_2, \cdots, x_ny_n\} \in F_q^n$. We will call $W_n = \{\boldsymbol{w}_1, \boldsymbol{w}_2, \cdots, \boldsymbol{w}_n\}$ over F_q^n an ordered basis of F_q^n if W_n is a basis of F_q^n and the ordering of n vectors $\boldsymbol{w}_1, \boldsymbol{w}_2, \cdots, \boldsymbol{w}_n$ has meaning. The subset W_i of W_n is defined by $W_i = \{\boldsymbol{w}_1, \boldsymbol{w}_2, \cdots, \boldsymbol{w}_i\}$ for $1 \le i \le n$.

Definition 2.1 For a vector $\boldsymbol{b} \in F_q^n$, the map of an ordered basis W_n denoted by $\sigma: F_q^n \to \{0,1,2, \cdots, n\}$ is defined as

$$\sigma(\boldsymbol{b}) = \min \{i \mid \boldsymbol{b} \in \mathrm{Span}\{W_i\}, 0 \le i \le n\},$$

where $\mathrm{Span}\{W_i\}$ is a subspace of F_q^n spanned by W_i and $\mathrm{Span}\{W_0\} = \{\boldsymbol{0}\}$.

Definition 2.2 Let $\boldsymbol{u}_i \in U_n$, $\boldsymbol{v}_j \in V_n$. The product $\boldsymbol{u}_i\boldsymbol{v}_j$ of $\boldsymbol{u}_i$ and $\boldsymbol{v}_j$ for an ordered basis W_n is said to be well-behaved if $\sigma(\boldsymbol{u}_u\boldsymbol{v}_v) < \sigma(\boldsymbol{u}_i\boldsymbol{v}_j)$ for any $1 \le u \le i$, $1 \le v \le j$, $(u, v) \ne (i, j)$.

Let W be a subset of an ordered basis W_n of F_q^n. The linear code $C(W_n, W)$ over F_q is defined as

$$C(W_n, W) = \mathrm{Span}\{W\}^\perp \subset F_q^n,$$

where $\mathrm{Span}\{W\}^\perp$ means the set of all vectors in F_q^n orthogonal to $\mathrm{Span}\{W\}$.

2.1 Miura's Definition of d_{FR}

Definition 2.3 For $1 \le s \le n$, we define $N(s)$ as

$$N(s) = \#\{(i,j) \mid \sigma(\boldsymbol{w}_i\boldsymbol{w}_j) = s,\ 1 \le i, j \le n,\ \boldsymbol{w}_i\boldsymbol{w}_j \text{ is well-behaved}\},$$

where $\#A$ means the cardinality of set A. For an ordered basis W_n of F_q^n we define $N(W_n)$ as $N(W_n) = (N(1), N(2), \cdots, N(n))$.

Lemma 2.1 [3] We have $1 \le N(s) \le s$, for $1 \le s \le n$.

Definition 2.4 The Feng-Rao designed minimum distance of the linear code $C(W_n, W)$ is denoted as $d_{FR}(C, W_n)$ and defined as

$$d_{FR}(C, W_n) = \min\{ N(s) \mid \boldsymbol{w}_s \in W_n \setminus W,\ 1 \le s \le n\},$$

where $W_n \setminus W$ is the subset of W_n without the elements of W.

We will use the notation $d_{FR}(C, W_n)$ to show the ordered basis W_n explicitly.

Lemma 2.2 [3] Let d be the true minimum distance of $C(W_n, W)$. Then we have $d \geq d_{FR}(C, W_n)$.

For a fixed linear code C we can choose many bases W_n such that $C = C(W_n, W)$.

Definition 2.5 For a linear code C we define the set of all ordered bases such that C can be defined by this ordered basis, i.e.,

$$\Re(C) = \{W_n \mid {}^{\exists}W \subseteq W_n \ \ s.t.\ C = C(W_n, W)\},$$

Note that W is uniquely determined from C and W_n. Our purpose is to give an optimum ordered basis W_n for a given linear code C.

Definition 2.6 The Feng-Rao designed minimum distance $d_{FR}(C)$ of C by Miura is defined as

$$d_{FR}(C) = \max\{\, d_{FR}(C, W_n) \mid W_n \in \Re(C)\}.$$

The ordered basis W_n satisfying $d_{FR}(C) = d_{FR}(C, W_n)$ is called an optimum ordered basis of C.

2.2 Properties of Miura's d_{FR}

Next we will consider only binary linear codes over F_2. The following lemma [3, Lemma 3.3] and its corollary [3, Corollary 3.4] are essential in our discussions. We will quote them in case of binary linear codes, although Miura proved them in case of linear codes over any F_q. We will give their proof for the convenience of readers who will find difficulty in obtaining Miura's thesis [3].

Lemma 2.3 Let $W_n = \{\boldsymbol{w}_1, \boldsymbol{w}_2, \cdots, \boldsymbol{w}_n\}$ be an ordered basis of $F_q^{\,n}$ and $\boldsymbol{w} \in F_2^{\,n}$. If $\sigma(\boldsymbol{w}_t\boldsymbol{w}) < t \leq n$, then there exists at least one i such that $\sigma(\boldsymbol{w}_t\boldsymbol{w}) \leq \sigma(\boldsymbol{w}_i\boldsymbol{w})$ and $1 \leq i < t$.

Proof We will show a contradiction if we assume

$$\sigma(\boldsymbol{w}_i\boldsymbol{w}) < \sigma(\boldsymbol{w}_t\boldsymbol{w}) \qquad \text{for} \quad i = 1,2,\cdots,t\text{-}1 \tag{1}$$

Let $\sigma(\boldsymbol{w}_t\boldsymbol{w}) = s < t$. We have

$$\boldsymbol{w}_t\boldsymbol{w} = \alpha_1\boldsymbol{w}_1 + \alpha_2\boldsymbol{w}_2 + \cdots + \alpha_s\boldsymbol{w}_s, \quad \alpha_s = 1. \tag{2}$$

$\alpha_i \in F_2$ for $1 \le i \le s$. From the assumption (1) we also have

$$\boldsymbol{w}_i\boldsymbol{w} = \beta_{i,1}\boldsymbol{w}_1 + \beta_{i,2}\boldsymbol{w}_2 + \cdots + \beta_{i,s-1}\boldsymbol{w}_{s-1}. \tag{3}$$

$\beta_{i,j} \in F_2$ for $1 \le i < t$, $1 \le j < s$. We have $\boldsymbol{ww} = \boldsymbol{w}$ for any binary vector $\boldsymbol{w}$ and we have

$$\begin{aligned} \boldsymbol{w}_t\boldsymbol{w} = \boldsymbol{w}_t\boldsymbol{ww} &= (\boldsymbol{w}_t\boldsymbol{w})\boldsymbol{w} = (\alpha_1\boldsymbol{w}_1 + \alpha_2\boldsymbol{w}_2 + \cdots + \alpha_s\boldsymbol{w}_s)\boldsymbol{w} \\ &= \alpha_1\boldsymbol{w}_1\boldsymbol{w} + \alpha_2\boldsymbol{w}_2\boldsymbol{w} + \cdots + \alpha_s\boldsymbol{w}_s\boldsymbol{w} \end{aligned} \tag{4}$$

from (2). However (2) and (4) are contradiction, since the term $\alpha_s\boldsymbol{w}_s$ in (2) is missing in the right-hand side of (4) because of (3). ♦

Corollary 2.1 If $\sigma(\boldsymbol{w}_i\boldsymbol{w}_j) = s$ and $\boldsymbol{w}_i\boldsymbol{w}_j$ is well-behaved, then we have $1 \le i, j \le s$.

Proof If we assume $i > s$, then application of Lemma 2.3 with $t = i$ and $\boldsymbol{w} = \boldsymbol{w}_j$ shows that $\boldsymbol{w}_i\boldsymbol{w}_j$ is not well-behaved from the definition of well-behavedness. The proof of $j \le s$ is the same. ♦

Therefore we will prove the following theorem shown in [5].

Theorem 2.1 Any binary linear code has $d_{FR}(C, W_n)$ equal to either one or an even number.

This theorem tells that binary linear codes can not have an odd $d_{FR}(C, W_n) \ge 3$. Our proof of Theorem 2.1 is very simple as shown below.
First we use the following property which is obvious from the definition.

Lemma 2.4 If $\boldsymbol{w}_i\boldsymbol{w}_j$ $(i \ne j)$ is well-behaved, then $\boldsymbol{w}_j\boldsymbol{w}_i$ is also well-behaved.

Next we use the following lemma which is almost obvious from Corollary 2.1.

Lemma 2.5 If $\boldsymbol{w}_i\boldsymbol{w}_i$ is well-behaved of W_n, then we have $N(i) = 1$.

Proof For a binary vector $\boldsymbol{w}_i$ we have $\boldsymbol{w}_i\boldsymbol{w}_i = \boldsymbol{w}_i$ and $\sigma(\boldsymbol{w}_i\boldsymbol{w}_i) = i$. There can not exist another $\boldsymbol{w}_u\boldsymbol{w}_v$ $(u, v) \ne (i, j)$ which satisfies the condition that $\sigma(\boldsymbol{w}_u\boldsymbol{w}_v) = i$ and $\boldsymbol{w}_u\boldsymbol{w}_v$ is well-behaved, since such $\boldsymbol{w}_u$ and $\boldsymbol{w}_v$ must satisfy $1 \le u, v \le i$

from Corollary 2.1 and $\sigma(w_u w_v)$ must be less than i because of $w_i w_i$ being well-behaved. ♦

Proof of Theorem 2.1 We have $N(1) = 1$ because of $\sigma(w_1 w_1) = 1$.

If $w_i w_i$ is not well-behaved for $i \geq 2$, then $N(i)$ is even for $i \geq 2$ from Lemma 2.5. Therefore we have Theorem 2.1.

If there exists such a $w_i \in W_n \setminus W$ as $w_i w_i$ is well-behaved, then we have $N(i) = 1$ from Lemma 2.5 and $d_{FR}(C, W_n) = 1$ from the definition of $d_{FR}(C, W_n)$. ♦

2.3 Matsumoto's Definition of d^*_{FR}

Miura's definition of the Feng-Rao designed minimum distance has been generalized to d^*_{FR} by Matsumoto and Miura [2].

Definition 2.7 For $1 \leq s \leq n$, we define $N^*(s)$ as

$$N^*(s) = \#\{(i, j) \mid \sigma(u_i v_j) = s, 1 \leq i, j \leq n,\ u_i v_j \text{ is well-behaved}\}.$$

For an ordered basis W_n, U_n and V_n of F_q^n we define $N^*(W_n, U_n, V_n)$ as $N^*(W_n, U_n, V_n) = (N^*(1), N^*(2), \cdots, N^*(n))$.

Our numerical experiments in case of all (7, k) linear codes by generating all the possible set of three bases show $N^*(1) \leq 1$ and $N^*(2) \leq 2$. So we have the following conjecture.

Conjecture 2.1 We have $0 \leq N^*(s) \leq s$, for $1 \leq s \leq n$.

Definition 2.8 The Feng-Rao designed minimum distance by Matsumoto of the linear code $C(W_n, W)$ is denoted as $d^*_{FR}(C, W_n, U_n, V_n)$ and defined as

$$d^*_{FR}(C, W_n, U_n, V_n) = \min\{ N^*(s) \mid w_i \in W_n \setminus W, 1 \leq s \leq n \}.$$

Lemma 2.6 [2] Let d be the true minimum distance of $C(W_n, W)$. Then we have $d \geq d^*_{FR}(C, W_n, U_n, V_n)$.

Definition 2.9 For a linear code C we define the set of all triples with three ordered bases such that C can be defined by an ordered basis W_n, i.e.,

$$\Im(C) = \left\{ (W_n, U_n, V_n) \,\middle|\, \begin{array}{l} {}^{\exists}W \subseteq W \text{ s.t. } C = C(W_n, W) \text{ and} \\ U_n, V_n \text{ are ordered basis of } F_q^n \end{array} \right\}.$$

Note that W is uniquely determined from C and W_n. Moreover U_n and V_n are not concerned with the linear code C. Our purpose is to give an optimum triple of three ordered bases (W_n, U_n, V_n) for a given linear code C.

Definition 2.10 The Feng-Rao designed minimum distance $d^*_{FR}(C)$ of C by Matsumoto is defined as

$$d^*_{FR}(C) = \max\{d^*_{FR}(C, W_n, U_n, V_n) \mid (W_n, U_n, V_n) \in \Im(C)\}.$$

The triple of three ordered basis (W_n, U_n, V_n) satisfying $d^*_{FR}(C) = d^*_{FR}(C, W_n, U_n, V_n)$ is called an optimum triple of C.

2.4 An Example of Matsumoto's d^*_{FR}

Next we will consider only binary linear codes over F_2. First, we show an example of a triple of three bases in order to discuss Matsumoto's generalization of d^*_{FR}.

Example For (7, 4) linear code, let three ordered bases be as follows:

$$\begin{aligned}
\boldsymbol{w}_1 = \boldsymbol{u}_1 &= (1\ 0\ 1\ 1\ 1\ 0\ 0), & \boldsymbol{v}_1 &= (0\ 0\ 0\ 0\ 0\ 1\ 0),\\
\boldsymbol{w}_2 = \boldsymbol{u}_2 &= (0\ 1\ 0\ 1\ 1\ 1\ 0), & \boldsymbol{v}_2 &= (0\ 0\ 0\ 0\ 0\ 1\ 1),\\
\boldsymbol{w}_3 = \boldsymbol{u}_3 &= (0\ 0\ 1\ 0\ 1\ 1\ 1), & \boldsymbol{v}_3 &= (1\ 0\ 1\ 1\ 1\ 0\ 0),\\
\boldsymbol{w}_4 = \boldsymbol{u}_4 &= (0\ 0\ 0\ 1\ 1\ 0\ 0), & \boldsymbol{v}_4 &= (0\ 0\ 0\ 1\ 1\ 0\ 0),\\
\boldsymbol{w}_5 = \boldsymbol{u}_5 &= (0\ 0\ 0\ 0\ 0\ 1\ 1), & \boldsymbol{v}_5 &= (0\ 1\ 0\ 1\ 1\ 1\ 0),\\
\boldsymbol{w}_6 = \boldsymbol{u}_6 &= (0\ 0\ 0\ 1\ 0\ 1\ 0), & \boldsymbol{v}_6 &= (0\ 0\ 1\ 0\ 1\ 1\ 1),\\
\boldsymbol{w}_7 = \boldsymbol{u}_7 &= (0\ 0\ 0\ 0\ 0\ 1\ 0), & \boldsymbol{v}_7 &= (0\ 0\ 0\ 1\ 0\ 1\ 0).
\end{aligned}$$

Since the matrix of $\sigma(\boldsymbol{u}_i\boldsymbol{v}_j)$ is

$$\begin{bmatrix}
0 & 0 & 1 & 4 & 4 & 5 & 7\\
7 & 7 & 4 & 4 & 2 & 6 & 6\\
7 & 5 & 5 & 7 & 6 & 3 & 7\\
0 & 0 & 4 & 4 & 4 & 7 & 7\\
7 & 5 & 0 & 0 & 7 & 5 & 7\\
7 & 7 & 7 & 7 & 6 & 7 & 6\\
7 & 7 & 0 & 0 & 7 & 7 & 7
\end{bmatrix}.$$

About matrix $\sigma(\boldsymbol{u}_i\boldsymbol{v}_j)$ is shows that $\sigma(\boldsymbol{u}_1\boldsymbol{v}_3) = 1$ and $\boldsymbol{u}_1\boldsymbol{v}_3$ is well-behaved, which contradicts with Corollary 2.1. Therefore in Matsumoto's definition we don't have Lemma 2.3 and Corollary 2.1, which are used in proving Theorem 2.1.

However our numerical experiments on (7, k) binary linear codes strongly suggest the following conjecture which is the same our previous Theorem 2.1.

Conjecture 2.2 Any binary linear code has $d^{*}_{FR}(C, W_n, U_n, V_n)$ equal to either one or an even number.

3. THE FENG-RAO DESIGNED MINIMUM DISTANCE OF CYCLIC CODES

This section is a first trial to find an optimum ordered basis W_n for (n, k) cyclic codes over F_q, which are in a class of the most useful linear codes. For a cyclic code C there exists a good designed minimum distance such as BCH designed minimum distance denoted by d_{BCH}. It is interesting to compare $d_{FR}(C)$ with d_{BCH} of a cyclic code C.

3.1 d_{FR} of Cyclic Codes and Type I Ordered Basis

In here the Feng-Rao designed minimum distance $d_{FR}(C)$ for (n, k) cyclic code C over F_q is investigated. We are interest in the relation between the choice of ordered basis $W_n = \{\boldsymbol{w}_1, \boldsymbol{w}_2, \cdots, \boldsymbol{w}_n\}$ and the Feng-Rao designed minimum distance $d_{FR}(C, W_n)$, since the value of $d_{FR}(C, W_n)$ depends on the choice of W_n [3]. If we define $d_{FR}(C)$ as the maximum value of $d_{FR}(C, W_n)$ among all the possible choices of W_n for C, it is believed that $d_{FR}(C)$ is a good lower bound of the minimum distance of C.

For a fixed linear code C we can choose many bases W_n such that $C = C(W_n, W)$.

Definition 3.1 For a cyclic code C we define the set $\aleph(C)$ of all ordered bases such that C can be defined by this ordered basis, i.e.,

$$\aleph(C) = \{W_n \mid {}^{\exists}W \subseteq W_n \ \ s.t.\ C = C(W_n, W)\}.$$

Note that W is uniquely determined from C and W_n. Our purpose is to give an optimum ordered basis W_n for a given cyclic code C.

Definition 3.2 The Feng-Rao designed minimum distance $d_{FR}(C)$ of C is defined as

$$d_{FR}(C) = \max\{\, d_{FR}(C, W_n) \mid W_n \in \aleph(C)\}.$$

The ordered basis W_n satisfying $d_{FR}(C) = d_{FR}(C, W_n)$ is called an optimum ordered basis for C.

The computation of $d_{FR}(C)$ needs the choice of an ordered basis of $F_q^{\,n}$, i.e.,

$$W_n = \{\boldsymbol{w}_1, \boldsymbol{w}_2, \cdots, \boldsymbol{w}_n\} \tag{5}$$

Now we will consider an (n, k) cyclic code C over a base field F_q. Let $g(x)$ be the generator polynomial of C and

$$h(x) = (x^n\text{-}1)/g(x) = x^k + w_2\, x^{k\text{-}1} + \cdots + w_{k+1} \tag{6}$$

be the check polynomial of C. It is well known that C consists of all vectors in $F_q^{\,n}$ orthogonal to

$$\boldsymbol{w} = (1, w_2, \cdots, w_{k+1}, 0, \cdots, 0) \in F_q^{\,n}, \qquad w_{k+1} \neq 0 \tag{7}$$

and $\delta(\boldsymbol{w}), \cdots, \delta^{n\text{-}k\text{-}1}(\boldsymbol{w})$, where $\delta(\boldsymbol{c}) = (c_n, c_1, \cdots, c_{n\text{-}1})$ for $\boldsymbol{c} = (c_1, c_2, \cdots, c_{n\text{-}1})$ is the right cyclic shift of $\boldsymbol{c}$. Therefore without loss of generality we will assume that

$$C = \mathrm{Span}\{\boldsymbol{w}, \delta(\boldsymbol{w}), \cdots, \delta^{n\text{-}k\text{-}1}(\boldsymbol{w})\}^{\perp}. \tag{8}$$

There are many choices for W_n. From Lemma 2.1 and Definition 2.4 we need to set $\boldsymbol{w}_s \in W$ for small s in order to obtain a large value of d_{FR}.

The following W_n is a natural choice for computing $d_{FR}(C, W_n)$ in case of (n, k) cyclic codes C.

Definition 3.3 The ordered basis W_n is called "Type I" if $W = \{\boldsymbol{w}_1, \boldsymbol{w}_2, \cdots, \boldsymbol{w}_{n\text{-}k}\}$ is a permutation of $\{\boldsymbol{w}, \delta(\boldsymbol{w}), \cdots, \delta^{n\text{-}k\text{-}1}(\boldsymbol{w})\}$, where $\boldsymbol{w}$ is the vector of (7).

3.2 d_{FR} of Nonbinary Cyclic Codes

First the following proposition will be proved about the product of two vectors of W.

Proposition 3.1 Let $W = \{\boldsymbol{w}_{u_1}, \boldsymbol{w}_{u_2}, \cdots, \boldsymbol{w}_{u_{n-k}}\}$ be a subset of an ordered basis W_n with Type I. For $1 \leq i \leq n\text{-}k$, $1 \leq j \leq n\text{-}k$, we have following two statements:

1. $w_{u_i} w_{u_i} = \alpha w_{u_i}$ with $\alpha \in F_q \setminus \{\mathbf{0}\}$ or $w_{u_i} w_{u_i} \notin \mathrm{Span}\{W\}$.
2. if $i \neq j$, then $w_{u_i} w_{u_j} = \mathbf{0}$ or $w_{u_i} w_{u_j} \notin \mathrm{Span}\{W\}$.

Proof Let $t_1 = w$ and $t_i = \delta^{i\text{-}1}(w)$ for i = 2,3,⋯,n-k. From Definition 3.3 we have

$$W = \{w_{u_1}, w_{u_2}, \cdots, w_{u_{n-k}}\} = \{t_1, t_2, \cdots, t_{n-k}\}.$$

We will first prove the statement 1. We have

$$t_1 t_1 = (1^2, \cdots, w_{k+1}^2, 0, \cdots, 0) \neq \mathbf{0} \tag{9}$$

When we assume $t_1 t_1$ = ∈ Span$\{W\}$, we have

$$t_1 t_1 = \sum_{i-1}^{n-k} \beta_i t_i \; .$$

In order to satisfy (9), we need $\beta_i = 0$ for $i = n-k, n-k-1$, ⋯, 2 from (7) and (9). Then we have $\beta_1 = \alpha \neq 0$ from $t_1 t_1 \neq \mathbf{0}$. Consequently we have $t_1 t_1 = \alpha\, t_1$. It is obvious that $t_1 t_1 \notin$ Span$\{W\}$ if $t_1 t_1 \neq \alpha\, t_1$.

The proof of $t_i t_i = \alpha\, t_i$ or $t_i t_i \notin$ Span$\{W\}$ is similar, since $t_i t_i = \alpha\, t_i$ is equivalent to $t_1 t_1 = \alpha\, t_1$ from $t_i = \delta^{i-1}(t_1)$ and $t_i t_i = \delta^{i-1}(t_1 t_1)$.

Next we will prove the statement 2. Since $t_i t_j = t_j t_i$ and

$$t_i t_j = \delta(t_{i-1} t_{j-1}) = \delta^{i-1}(t_1 t_{j-i+1})$$

for $i < j$, it is enough to prove the case of $i = 1$ and $2 \leq j \leq n-k$. We have

$$t_1 t_j = (0^{j-1}, *, \cdots, *, 0^{n-k-1}), \tag{10}$$

where 0^u denotes the zero vector of length u and $*$ means any value in F_q. When we assume $t_1 t_j \in \mathrm{Span}\{W\}$, we have

$$t_1 t_j = \sum_{i=1}^{n-k} \gamma_i t_i .$$

From (6) and (10) we have $t_1 t_j = \mathbf{0}$. ♦

The following proposition shows that the choice of an ordered basis W_n with Type I is worst in many cases of nonbinary cyclic codes.

Proposition 3.2 Let C be a nonbinary (n, k) cyclic code. We have

$$d_{FR}(C, W_n) \le 1, \tag{11}$$

if W_n is Type I and the check polynomial $h(x)$ of (3.2) has a coefficient $b_i \neq 0, 1$.

Proof If $\boldsymbol{w}_1 \notin W$ we have $d_{FR}(C, W_n) \le 1$ from Lemma 2.1 and Definition 2.4. Then we consider the case of $\boldsymbol{w}_1 \in W$. From Proposition 3.1,

$$\boldsymbol{w}_1 \boldsymbol{w}_1 = (0, \cdots, 0, 1^2, w_2^2, \cdots, w_{k+1}^2, 0, \cdots, 0) = \alpha\, \boldsymbol{w}_1, \quad \alpha \in F_q \setminus \{0\} \tag{12}$$

or

$$\boldsymbol{w}_1 \boldsymbol{w}_1 \notin \mathrm{Span}\{W\}. \tag{13}$$

From (12) we need $w_i = 0$ or α for all $i = 1, 2, \cdots, k+1$. Since $w_i = 1 \neq 0$, we have $\alpha = 1$. If there exists a $w_i \neq 0, 1$ with $i = 2, 3, \cdots, k+1$, then (12) can not hold.

In case of (13), we have $\sigma(\boldsymbol{w}_1 \boldsymbol{w}_1) = r > 1$ and $N(r) = 1$, which means $d_{FR}(C, W_n) \le 1$ because of $\boldsymbol{w}_r \in W_n \setminus W$. ♦

Proposition 3.2 shows that the choice of any Type I ordered basis is not good for most of nonbinary cyclic codes, although no efficient method has been known for computing good ordered basis.

3.4 d_{FR} of Binary Cyclic Codes

We consider (n, k) cyclic codes over F_2 with an odd length n and $k = 1$, $k = 2$, and $k = n - 1$ in this section. We will show that there exists not so bad ordered basis with Type I.

3.4.1 $(n, 1)$ Binary Cyclic Code (Repetition Code)

In this case the code has only two code words, i.e., $\mathbf{0} = (0, 0, \cdots, 0)$ and $\mathbf{1} = (1, 1, \cdots, 1)$. Since its generator polynomial $g(x) = x^{n-1} + x^{n-2} + \cdots + x + 1$ has n-1 roots, i.e., $\alpha, \alpha^2, \cdots, \alpha^{n-1}$ with α being a primitive n-th root of unity, we have

$$d_{BCH} = n. \tag{14}$$

Lemma 3.1 For a binary repetition code C ($k = 1$), we have $d_{FR}(C, W_n) \leq n-1$.

Proof If $w_n \in W$, then we have $W_n \setminus W = \{w_s\}$ $(1 \leq s \leq n-1)$ and $d_{FR}(C, W_n) = N(s) \leq n-1$ from Lemma 2.1.

Therefore we consider the case where $W_n \setminus W = \{w_n\}$ and $d_{FR}(C, W_n) = N(n)$. We have $N(n) = n$ if and only if

$$\begin{cases} \sigma(w_i w_j) < n & \text{if } i + j < n+1, \\ \sigma(w_i w_{n+1-i}) = n & \text{if } 1 \leq i \leq n. \end{cases} \tag{15}$$

For a binary vector w_i we have

$$w_i w_i = w_i \text{ and } \sigma(w_{\frac{n+1}{2}} w_{\frac{n+1}{2}}) = \frac{n+1}{2} < n.$$

Therefore (15) can not hold. ◆

Proposition 3.3 For a binary $(n, 1)$ cyclic code C, we have $d_{FR}(C, W_n) = n-1$.

Proof We will show the existence of a Type I ordered basis $W_n = \{w_1, w_2, \cdots, w_n\}$ which satisfies $d_{FR}(C, W_n) = n-1$. Let

$$e_i = (0^{i-1}, 1, 0^{n-i}) \in F_2^n, \tag{16}$$

From $h(x) = (x^n - 1)/g(x) = x + 1$, we can choose W_n as

$$\begin{cases} w_i = e_{2i-1} + e_{2i} & \text{for } 1 \leq i \leq \frac{n-1}{2}, \\ w_{\frac{n-1}{2}+i} = e_{n-(2i-2)} + e_{n-(2i-1)} & \text{for } 1 \leq i \leq \frac{n-1}{2}, \\ w_n = e_1, \end{cases}$$

and $W = W_{n-1}$. Since $e_i \notin \text{Span}\{W_{n-1}\}$ for $i = 1, 2, \cdots, n$, we have

$$\begin{cases} \sigma(w_i w_i) = i & \text{if } 1 \le i \le n, \\ \sigma(w_i w_j) = 0 & \text{if } 1 \le i < j \le n-1-i, \\ \sigma(w_i w_{n-i}) = n & \text{if } 1 \le i \le n-1. \end{cases} \tag{17}$$

We have $N(n) \ge n-1$ and $d_{FR}(C, W_n) \ge n-1$. From Lemma 3.1 we have $d_{FR}(C, W_n) = n-1$. ◆

Therefore from (14) and Proposition 3.3, we have $\lfloor \frac{d_{FR}-1}{2} \rfloor = \lfloor \frac{d_{BCH}-1}{2} \rfloor - 1$.

3.4.2 (*n*, 2) Binary Cyclic Code

Let C be a binary $(n, 2)$ cyclic code. In this case, it is obvious that the check polynomial of C is only $h(x) = x^2 + x + 1$. We have $x^3 - 1 | x^n - 1$ and $3 | n$, since $h(x) | x^n - 1$, $x - 1 | x^n - 1$, $x^3 - 1 = (x-1)h(x)$, and

$$x^n - 1 = x^r(x^3 - 1)(x^{3(m-1)} + x^{3(m-2)} + \cdots + x^3 + 1) + (x^r - 1)$$

for $n = 3m + r$ and $0 \le r \le 2$.

Lemma 3.2 The odd code length of binary $(n, 2)$ cyclic codes C is a multiple of 3.

Let α be a primitive n-th root of unity, then two roots of the check polynomial of C are $\alpha^{n/3}$ and $\alpha^{2n/3} = \alpha^{-n/3}$. Since the number of consecutive roots of the generator polynomial of C is $\frac{2}{3}n - 1$, we have

$$d_{BCH} = \frac{2}{3}n. \tag{18}$$

Proposition 3.4 For a binary $(n, 2)$ cyclic code C, we have

$$d_{FR}(C, W_n) \ge 2(\frac{n}{3} - 1) \tag{19}$$

with a Type I ordered basis W_n.

Proof A lower bound of (19) is shown by finding an ordered basis W_n satisfying equation (19). Let $W_n = \{w_1, w_2, \cdots, w_n\}$ be an ordered basis with

$$\begin{cases} w_i = e_{3i-2} + e_{3i-1} + e_{3i} & \text{for } 1 \le i \le \frac{n}{3}, \\ w_{\frac{n}{3}+2k-1} = e_{n-3k} + e_{n-3k+1} + e_{n-3k+2} & \text{for } 1 \le k \le \frac{n}{3}-1, \\ w_{\frac{n}{3}+2k} = e_{n-3k-1} + e_{n-3k} + e_{n-3k+1} & \text{for } 1 \le k \le \frac{n}{3}-1, \\ w_{n-1} = e_3,\ w_n = e_2, & \end{cases}$$

where e_i is defined by (16). From this we have $W = W_{n-2}$ and $d_{FR}(C, W_n) = \min\{N(n-1), N(n)\}$. We can writer e_{3i}, e_{3i-1} and e_{3i-2} for $1 \le i \le \frac{n}{3}$ as

$$e_{3i} = w_{n-1} + (\#),\quad e_{3i-1} = w_n + (\#),\quad e_{3i-2} = w_n + w_{n-1} + (\#),$$

where (#) represents a vector in Span{ W_{n-2}}. Therefore we have

$$\begin{cases} \sigma(w_i w_j) = 0 & \text{if } i+1 \le j < n-2i-1, \\ \sigma(w_i w_{n-2i-1}) = n-1, & \\ \sigma(w_i w_{n-2i}) = n & \end{cases} \tag{20}$$

for $1 \le i \le \frac{n}{3}-1$, since $w_i w_j = \mathbf{0}$ for $i+1 \le j < n-2i-1$, $w_i w_{n-2i-1} = e_{3i}$, and $w_i w_{n-2i} = e_{3i} + e_{3i-1}$.

Hence we have $N(n-1) = N(n) = 2(\frac{n}{3}-1)$ and $d_{FR}(C, W_n) = 2(\frac{n}{3}-1)$. ♦

Although Proposition 3.4 gives a lower bound of $d_{FR}(C, W_n)$ for a binary $(n,\ 2)$ cyclic code C with a Type I ordered basis W_n, the following consideration suggests us to conjecture that the lower bound is also an upper bound. Let $w_i' = \delta^{i-1}(w)$ $(1 \le i \le 7)$, i.e.,

$$\begin{aligned} w_1' &= (1\ \ 1\ \ 1\ \ 0\ \ 0\ \ 0\ \ 0\ \ 0\ \ 0), \\ w_2' &= (0\ \ 1\ \ 1\ \ 1\ \ 0\ \ 0\ \ 0\ \ 0\ \ 0), \\ w_3' &= (0\ \ 0\ \ 1\ \ 1\ \ 1\ \ 0\ \ 0\ \ 0\ \ 0), \\ w_4' &= (0\ \ 0\ \ 0\ \ 1\ \ 1\ \ 1\ \ 0\ \ 0\ \ 0), \\ w_5' &= (0\ \ 0\ \ 0\ \ 0\ \ 1\ \ 1\ \ 1\ \ 0\ \ 0), \\ w_6' &= (0\ \ 0\ \ 0\ \ 0\ \ 0\ \ 1\ \ 1\ \ 1\ \ 0), \\ w_7' &= (0\ \ 0\ \ 0\ \ 0\ \ 0\ \ 0\ \ 1\ \ 1\ \ 1). \end{aligned}$$

From this for each $i\,(1 \le i \le 7)$ the number of the pair $\{w_i', w_j'\}\,(i \ne j)$ satisfying $w_i' w_j' = \mathbf{0}$ $(1 \le j \le 7)$ is equal to 2 (i = 1, 7), 3 (i = 2, 6), or 4 (i = 3, 4, 5). For such a pair we have

$$\mathbf{0} \neq \boldsymbol{w}_i^{'} \boldsymbol{w}_j^{'} \in W_n \setminus W = \{\boldsymbol{w}_8, \boldsymbol{w}_9\}$$

from Proposition 3.1. Therefore we can make $d_{FR}(C\ W_n) = 6 > 4$, if $\sigma(\boldsymbol{w}_i \boldsymbol{w}_j)$ is the form

$$\begin{bmatrix} 1000089{*}{*} \\ 020089{*}{*}{*} \\ 00389{*}{*}{*}{*} \\ 0084{*}{*}{*}{*}{*} \\ 089{*}5{*}{*}{*}{*} \\ 89{*}{*}{*}6{*}{*}{*} \\ 9{*}{*}{*}{*}{*}7{*}{*} \\ {*}{*}{*}{*}{*}{*}{*}8{*} \\ {*}{*}{*}{*}{*}{*}{*}{*}9 \end{bmatrix}, \qquad (21)$$

which gives $N(8) = N(9) = 6$.

The first row of (21) requests that $\boldsymbol{w}_1$ should be $\boldsymbol{w}_1^{'}$ or $\boldsymbol{w}_7^{'}$. We consider the case of $\boldsymbol{w}_1 = \boldsymbol{w}_1^{'}$. The case of $\boldsymbol{w}_1 = \boldsymbol{w}_7^{'}$ is similar.

From the form of the first row of (21), we have $\boldsymbol{w}_6 = \boldsymbol{w}_2^{'}$, $\boldsymbol{w}_7 = \boldsymbol{w}_3^{'}$ or $\boldsymbol{w}_6 = \boldsymbol{w}_3^{'}$, $\boldsymbol{w}_7 = \boldsymbol{w}_2^{'}$. However any choice of $\boldsymbol{w}_2 = \{\boldsymbol{w}_4^{'}, \boldsymbol{w}_5^{'}, \boldsymbol{w}_6^{'}, \boldsymbol{w}_7^{'}\}$ can not satisfy the form of the second row of (21). Therefore (21) is impossible. From the symmetric property $\sigma(\boldsymbol{w}_i \boldsymbol{w}_j) = \sigma(\boldsymbol{w}_j \boldsymbol{w}_i)$ and $\sigma(\boldsymbol{w}_i \boldsymbol{w}_i) = i$ $(1 \le i \le n)$ for a binary codes, the value of $d_{FR}(C, W_n)$ is even if $d_{FR}(C, W_n) > 1$ and so $d_{FR}(C, W_n) < 6$ means $d_{FR}(C, W_n) \le 4$.

From the above discussion we strongly conjecture that $d_{FR}(C, W_n) = 2(\frac{n}{3} - 1) < d_{BCH}$ and $\left\lfloor \frac{d_{FR}-1}{2} \right\rfloor = \left\lfloor \frac{d_{BCH}-1}{2} \right\rfloor - 1$ with a Type I ordered basis W_n.

3.4.3 (*n*, *n*-1) Binary Cyclic Code (Parity Code)

Let C be a parity (n, n-1) code with $h(x) = x^{n-1} + x^{n-2} + \cdots + x + 1$. From $\boldsymbol{w}_1 = (1, 1, \cdots, 1) \in F_2^n$, we have $\sigma(\boldsymbol{w}_1 \boldsymbol{w}_j) = j$ for $j = 1, 2, \cdots, n$ and $N(2) = 2$. We have $d_{FR}(C, W_n) = 2$, since $W = \{\boldsymbol{w}_1\}$ and $W_n \setminus W = \{\boldsymbol{w}_2, \boldsymbol{w}_3, \cdots, \boldsymbol{w}_n\}$. Since the minimum distance of the parity code is 2, we have following proposition.

Proposition 3.5 For the parity (n, n-1) code C, we have $d_{FR}(C, W_n) = d_{BCH} = d = 2$.

4. CONCLUSION

In this Chapter it is proved that $d_{FR}(C,W_n)$ of binary linear codes can not take an odd number greater than or equal to 3 if we use Miura's definition [3] of $d_{FR}(C,W_n)$ for binary linear codes. We discussed Miura's definition and Masumoto's definition of the Feng-Rao designed minimum distance for binary linear codes. Matsumoto's definition is a generalization of Miura's definition. Some properties and examples induce some conjectures which tell Matsumoto's generalization is not so effective compared with Miura's definition for binary linear codes. Finally, it is shown that the choice of an ordered basis W_n with Type I is worst in many cases of nonbinary cyclic codes, since $d_{FR}(C,W_n) \leq 1$ if the check polynomial $h(x)$ has a coefficient neither equal to 0 nor 1. It was also shown that in case of binary (n, k) cyclic codes C with k=1, 2 and n-1, there exists an ordered basis W_n with Type I such that $d_{FR}(C,W_n) = n-1$, $d_{FR}(C,W_n) = 2(\frac{n}{3}-1)$, and $d_{FR}(C,W_n) = 2$ for k=1, 2, and n-1, respectively.

Finding an optimum ordered basis W_n for computing $d_{FR}(C)$ of arbitrary (n, k) cyclic code C or arbitrary linear code C is a desirable future work. Further investigation of the computational complexity of the Feng-Rao decoding of linear codes and their modification for cyclic codes are very important future problems.

REFERENCES

[1] G.L. Feng, T.R.N. Rao, "Decoding algebraic geometric codes up to the designed minimum distance," *IEEE Trans. Inform. Theory*, vol. 39, pp. 36-47, Jan. 1993.

[2] R. Matsumoto and S. Miura, "On the Feng-Rao Bound for the *L*-construction of Algebraic Geometric Codes," *IEICE Trans. Fundamentals*, vol. E83-A, No.5, pp. 923-926, May 2000.

[3] S. Miura, *Study on Error Correcting Codes Based on Algebraic Geometry*, Ph.D. Dissertation, the University of Tokyo, 1997. (in Japanese)

[4] J. Zheng, T. Kaida, K. Imamura, "A note on Feng-Rao designed minimum distance for cyclic codes," *The Third International Conference on Information, Communications & Signal Processing (ICICS 2001)*, pp. 1-5 (1AI.2) Oct., 2001.

[5] J. Zheng, T. Kaida, K. Imamura, "The Feng-Rao designed minimum distance of binary linear codes does not have an odd number except one," *The First International Workshop on Sequence Designed and Applications for CDMA Systems (IWSDA 2001)*, pp. 146-149, Sep., 2001.

Chapter 14

ON A USE OF GOLAY SEQUENCES FOR ASYNCHRONOUS DS CDMA APPLICATIONS

Jennifer R. Seberry, Beata J. Wysocki, and Tadeusz A. Wysocki
University of Wollongong, NSW 2522, Australia

Abstract: Golay complementary sequences, often referred to as Golay pairs, are characterized by the property that the sum of their aperiodic autocorrelation functions equals to zero, except for the zero shift. Because of this property, Golay complementary sequences can be utilized to construct Hadamard matrices defining sets of orthogonal spreading sequences for DS CDMA systems of the lengths not necessary being a power of 2. In the paper, we present an evaluation, from the viewpoint of asynchronous DS CDMA applications, of some sets of spreading sequences derived from Golay complementary sequences. We then modify those sets of sequences to enhance their correlation properties for asynchronous operation and simulate a multi-user DS CDMA system utilizing the modified sequences.

Key words: Spread Spectrum, Spreading Sequences, Correlation Functions, Golay Complementary Sequences

1. INTRODUCTION

Orthogonal spreading sequences are used in direct sequence code division multiple access (DS CDMA) systems for channel separation and to provide a spreading gain, e.g. [1]. The most popular class of such spreading sequences is the sets of Walsh-Hadamard sequences [2], which are easy to generate. However, the cross-correlation between two Walsh-Hadamard sequences can rise considerably in magnitude if there is a non-zero delay shift between them. Unfortunately, this is very often the case for an up-link (mobile to base station) transmission, due to differences in the corresponding propagation delays. As a result, significant multi-access interference (MAI)

[3] occurs which needs to be combated either by complicated multi-user detection algorithms [4], or reduction in bandwidth utilization. Moreover, due to their very regular structure, Walsh-Hadamard sequences are characterized with very poor auto-correlation properties. In real systems, this is alleviated by the use of scrambling codes on the top of Walsh-Hadamard sequences. These are normally very long codes having very distinctive peaks at zero in their auto-correlation functions. For example, in UMTS these cods are 2^{18} bits long [5]. In addition to improving synchronization properties, scrambling also helps in reducing MAI.

Another important drawback of using Walsh-Hadamard sequences or modified Walsh-Hadamard sequences [6] is the fact that sequence length must be equal to the integer power of 2. This is not the case if orthogonal sequences based on complementary Golay sequences are used instead of Walsh-Hadamard sequences.

Complementary pairs and sets of orthogonal spreading sequences defined on the base of Golay complementary sequences have been long studied for application in DS CDMA systems, [7], [8], [9]. In this paper, we want to extend those considerations to the sequences based on Golay-Hadamard matrices modified using the method introduced in [6]. As a result of modification, the new sets of orthogonal sequences are characterized with much lower peaks in aperiodic cross-correlation functions, and exhibit good autocorrelation properties, too. To illustrate usefulness of the modified sequences, we then simulate the multi channel DS CDMA system utilizing those modified sequences and compare the results to those obtained in the case of unmodified sequences. Our considerations are not limited to bipolar sequences but we investigate quadri-phase sequences as well.

The chapter is organized as follows. In the next section, we show some basic techniques to design Hadamard matrices based on complementary Golay sequences. In Section 3, we describe the sequence modification method, and show the correlation parameters for some sequence sets of lengths 26 and 32. The simulation of DS CDMA system utilizing the developed sequences is presented in Section 4, and Section 5 concludes the chapter.

2. CONSTRUCTION OF HADAMARD MATRICES USING GOLAY SEQUENCES

For a pair of Golay complementary sequences S_1, and S_2, the sum of their aperiodic autocorrelation functions equals to zero, except for the zero shift [10]:

$$c_{S_1}(\tau)+c_{S_2}(\tau)=0 \qquad \tau \neq 0 \tag{1}$$

where $c_{S_1}(\tau)$ and $c_{S_2}(\tau)$ denote the aperiodic autocorrelation functions [10].

It can be proven [11], [12] that if matrices **A** and **B** are the circulant matrices created from a pair of Golay complementary sequences S_1, and S_2 then the matrix

$$\mathbf{G}=\begin{bmatrix} \mathbf{A} & \mathbf{B} \\ \mathbf{B}^{\mathrm{T}} & -\mathbf{A}^{\mathrm{T}} \end{bmatrix} \tag{2}$$

is a Hadamard matrix. The matrix **G** can be used to generate spreading sequences for DS CDMA applications.

Originally, Golay complementary sequences have been defined as bipolar sequences, and one can generate bipolar Golay sequences of all lengths N, such that

$$N=2^a 10^b 26^c \tag{3}$$

where a, b, c, are non-negative integers. Hence, using the construction prescribed by Eq. (2), the sets of orthogonal spreading sequences of length $2N$ can be designed.

Another technique, which can be employed to construct Hadamard matrices from bipolar complementary sequences, is based on the Goethals-Seidel array [13]. In [14], Craigen, Holzmann and Kharaghani have shown that if U and V are Golay complementary sequences of length l_1 and X, Y are Golay complementary sequences of length l_2 then

$$\begin{aligned} A&=[U,X] \\ B&=[U,-X] \\ C&=[V,Y] \\ D&=[V,-Y] \end{aligned}$$

are four complementary sequences of length l_1+l_2, and that there is a Hadamard matrix of order

$$N=4(l_1+l_2)$$

constructed from Goethals-Seidel array. This technique produces many more Hadamard matrices than the construction given by Eq. (2).

However, much more Hadamard matrices can be produced if instead of bipolar Golay sequences, complementary quadri-phase Golay sequences are used [15]. The quadri-phase or complex Golay complementary sequences can be used to create quadri-phase Hadamard matrices and the resulting quadri-phase spreading sequences can be easily used in DS CDMA systems.

Holzmann and Kharaghani have published in [16] the results of a computer search for quadri-phase Golay sequences of lengths up to 13, and in [14] a methodology for designing of quadri-phase Golay sequences is given.

To show a usefulness of complementary Golay sequences for asynchronous DS CDMA systems, let us consider here the correlation properties of spreading sequence sets defined by Hadamard matrices derived from the complementary Golay sequences of length 13 and 16.

EXAMPLE 1

In [16], two quadri-phase complementary Golay sequences of length 13 are presented. These are:

$$S_{1\text{-}13} = [\ 1\ 1\ 1\ j\ -1\ 1\ 1\ -j\ 1\ -1\ 1\ -j\ j\ ]$$
$$S_{2\text{-}13} = [\ 1\ j\ -1\ -1\ -1\ j\ -1\ 1\ 1\ -j\ -1\ 1\ -j\ ]$$

To construct the set of 26-chip spreading sequences, we first obtain two circulant matrices of order 13, **A** and **B,** and then substitute them into Eq. (2) to obtain the Hadamard matrix of order 26.

The resulting 26-chip Golay-Hadamard sequences are characterized by the following correlation parameters:

$$C_{\max} = 0.9615$$
$$R_{CC} = 0.9777$$
$$R_{AC} = 0.5574$$

where $C_{\max}$ denotes the maximum value of a peak in aperiodic cross-correlation functions between any pair of the sequences, R_{CC} denotes the mean square aperiodic cross-correlation for the set of sequences [17], and R_{AC} denotes the mean square aperiodic autocorrelation value for the sequence set.

EXAMPLE 2

Another tested set of is obtained from two 16-chip bipolar Golay complementary sequences, $S_{1\text{-}16}$ and $S_{2\text{-}16}$, also given in [16]:

$$S_{1\text{-}16} = [++---+-+++++-++-]$$
$$S_{2\text{-}16} = [++---+-+----+--+]$$

The resulting set of 32 Golay-Hadamard sequences of length 32 has the following parrameters:

$$C_{\max} = 0.9688$$
$$R_{CC} = 0.9849$$
$$R_{AC} = 0.4688$$

From both examples it is clearly visible, due to the low values of R_{AC} that the resulting sets of spreading sequences have both very good synchronization properties, much better than it is in the case of Walsh-Hadamard sequences, where for the length of 32 we have:

$$C_{\max} = 0.9688$$
$$R_{CC} = 0.7873\ .$$
$$R_{AC} = 6.5938$$

3. SEQUENCE MODIFICATION METHOD

Further improvement to the values of correlation parameters of the sequence sets based on Golay-Hadamard matrices, can be obtained using the method introduced in [6]. That method is based on the fact that for a matrix **H** to be orthogonal, it must fulfill the condition

$$\mathbf{H}\mathbf{H}^T = N\mathbf{I} \tag{4}$$

where $\mathbf{H}^T$ is the transposed Hadamard matrix of order N, and $\mathbf{I}$ is the $N \times N$ unity matrix. The modification is achieved by taking another orthogonal $N \times N$ matrix $\mathbf{D}_N$, and the new set of sequences is based on a matrix $\mathbf{W}_N$, given by:

$$\mathbf{W}_N = \mathbf{H}\mathbf{D}_N \tag{5}$$

The matrix $\mathbf{W}_N$ is also orthogonal, since:

$$\mathbf{W}_N\mathbf{W}_N^T = \mathbf{H}\mathbf{D}_N(\mathbf{H}\mathbf{D}_N)^T = \mathbf{H}\mathbf{D}_N\mathbf{D}_N^T\mathbf{H}^T \tag{6}$$

and because of the orthogonality of matrix $\mathbf{D}_N$, we have

$$\mathbf{D}_N \mathbf{D}_N^T = k\mathbf{I} \tag{7}$$

where k is a real constant. Substituting (7) into (6) yields

$$\mathbf{W}_N \mathbf{W}_N^T = k\mathbf{H}\mathbf{H}^T = kN\mathbf{I}\,. \tag{8}$$

In addition, if $k = 1$, then the sequences defined by the matrix $\mathbf{W}_N$ are not only orthogonal, but possess the same normalization as the sequences defined by the original Hadamard matrix $\mathbf{H}$. However, other correlation properties of the sequences defined by $\mathbf{W}_N$ can be significantly different to those of the original sequences.

From equation (5) it is not clear how to chose the matrix $\mathbf{D}_N$ to achieve the desired properties of the sequences defined by the $\mathbf{W}_N$. In addition, there are only a few known methods to construct the orthogonal matrices. However, a simple class of orthogonal matrices of any order are diagonal matrices with their elements $d_{i,j}$ fulfilling the condition:

$$\left|d_{m,n}\right| = \begin{cases} 0 \ \mathtt{for}\ m \neq n \\ k \ \mathtt{for}\ m = n \end{cases}; \qquad m, n = 1, \ldots, N \tag{9}$$

To preserve the normalization of the sequences, the elements of $\mathbf{D}_N$, being in general complex numbers, must be of the form:

$$d_{m,n} = \begin{cases} 0 & \mathtt{for}\ m \neq n \\ \exp(j\phi_m) & \mathtt{for}\ m = n \end{cases}; \qquad m, n = 1, \ldots, N \tag{10}$$

where the phase coefficients ϕ_m; $m = 1, 2, \ldots, N$, are real numbers taking their values from the interval $[0, 2\pi)$. The values of ϕ_m; $m = 1, 2, \ldots, N$, can be optimized to achieve the desired correlation and/or spectral properties, e.g. minimum out-off-phase autocorrelation or minimal value of peaks in aperiodic cross-correlation functions.

From the application point of view, the most important classes of spreading sequences are bipolar sequences and quadri-phase sequences. In the first case, the values of ϕ_m; $m = 1, 2, \ldots, N$, are limited to $\{0, \pi\}$, and in the second case the coefficients ϕ_m can take values from the set $\{0, 0.5\pi, \pi, 1.5\pi\}$.

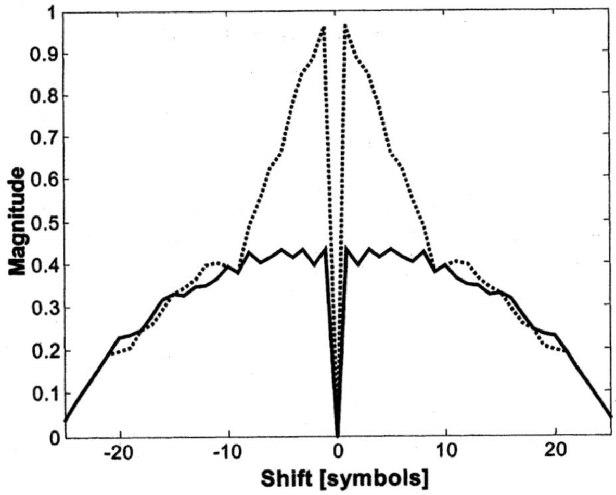

Figure 1. Peak magnitude of aperiodic cross-correlation functions for the families of spreading sequences; dotted line – sequences defined by original Golay-Hadamard matrix, solid line - modified sequences

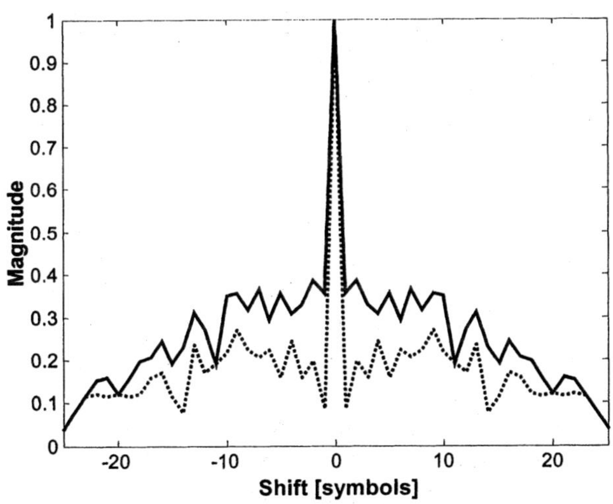

Figure 2. Peak magnitude of aperiodic auto-correlation functions for the families of spreading sequences; dotted line – sequences defined by original Golay-Hadamard matrix, solid line - modified sequences

We now show the results of applying the diagonal modification method to improve correlation properties of sequence sets constructed in the previous section. As an exhaustive search, in both cases, would require too much of processing time, we have performed a random search of 1000 trials, and searched for the diagonal matrix producing the lowest value of C_{max} for the modified sequence set.

EXAMPLE 3

In this example we show the results of minimizing C_{max} for the 26-chip sequence set developed in Example 1. After 10000 trials, the lowest value of C_{max} has been achieved for the matrix D_{26} with the following quadri-phase symbols on the diagonal:

$$[1 1 j -1 -j -j -j j 1 -1 j j 1 1 -1 -1 -1 -1 -1 1 1 1 j -1 -1 1].$$

The modified sequences have the following correlation parameters:

$$C_{max} = 0.4300$$
$$R_{CC} = 0.9599$$
$$R_{AC} = 1.0016$$

In Fig.1 and Fig.2, we present the plots of the maximum peak magnitudes for the cross-correlation functions between any possible pair of the modified sequence set, and the maximum peaks in the auto-correlation functions for all the modified sequences, respectively. For the comparison, we also present there the corresponding plots for the unmodified sequence set.

EXAMPLE 4

In this example, we show the results of minimizing C_{max} for the 32-chip sequence set developed in Example 1. Because the original sequences are bipolar, we first searched for the modification matrix D_{32} with the bipolar diagonal elements. After 10000 trials, the lowest value of C_{max} has been achieved in 9 different cases. The corresponding diagonals and the resulting correlation parameters are listed in Table 1.

In Fig. 3 and 4, we present the plots of the maximum peak magnitudes for the cross-correlation functions between any possible pair of the modified sequence set, and the maximum peaks in the auto-correlation functions for all the modified sequences, respectively. These modified sequences have been obtained using D_{32} with the diagonal:

[- - + - + - - - - + - + + - + - - + - + - + + + + - + - + - + +]

Table 1. The diagonals of D_{32} and the corresponding correlation parameters

Diagonal elements	C_{max}	R_{CC}	R_{AC}
+++---++-----++++-+--+++--+++-+-	0.4688	0.9638	1.1211
++--++---+++-++---+---+-+--+++++	0.4688	0.9724	0.8555
+++--++-----++----++--+-+++--+-+	0.4688	0.9735	0.8203
--+++++-+-++++-+-+--+--+-++-+---	0.4688	0.9711	0.8945
++--+--+++++++-----+-++++---++++	0.4688	0.9676	1.0039
-+++--+--+++-+--+---+--+----++-+	0.4688	0.9681	0.9883
-+---+-++-++--+-+-+--++-++-+++-+	0.4688	0.9676	1.0039
-+-+-+++--+++-+--++-+++-++-+--++	0.4688	0.9675	1.0078
--+-+----+-++-+--+-+-++++-+-+-++	0.4688	0.9594	1.2578

Even better results can be obtained if the modifying matrix D_{32} has the quadri-phase diagonal elements. After 10000 random trials, the diagonal leading to the lowest value of C_{max} has been found to be:

[-1-1j-jj-j-j-jj11-11-1j-1-11j1-1j-1-1-j11-j-j-j-j1],

The following parameters of the modified sequence set have been achieved:

$$C_{max} = 0.3763$$
$$R_{CC} = 0.9681$$
$$R_{AC} = 0.9902$$

In Fig. 5 and 6, we present the plots of the maximum peak magnitudes for the cross-correlation functions between any possible pair of the modified sequence set, and the maximum peaks in the auto-correlation functions for all the modified sequences, respectively.

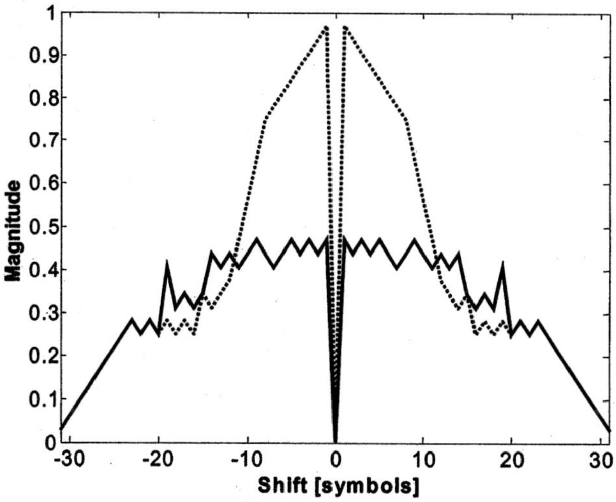

Figure 3. Peak magnitude of aperiodic cross-correlation functions for the families of spreading sequences; dotted line – sequences defined by original Golay-Hadamard matrix, solid line - modified bipolar sequences

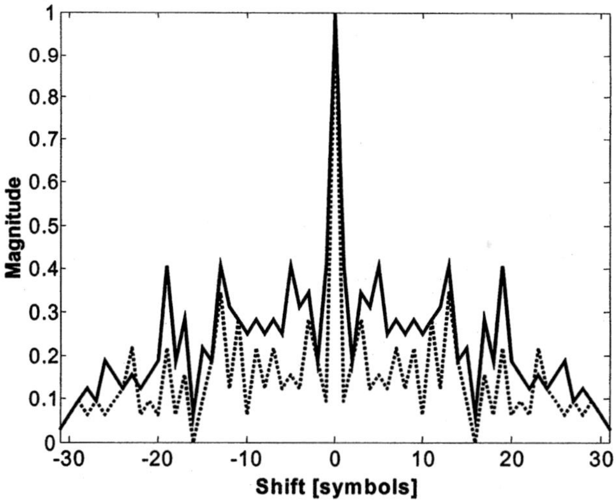

Figure 4. Peak magnitude of aperiodic auto-correlation functions for the families of spreading sequences; dotted line – sequences defined by original Golay-Hadamard matrix, solid line - modified bipolar sequences

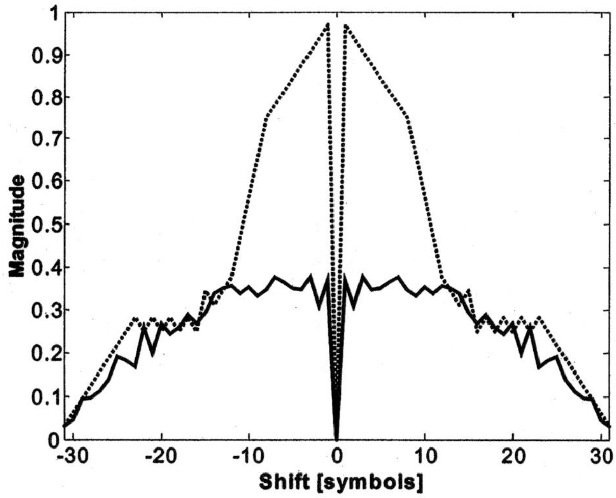

Figure 5. Peak magnitude of aperiodic cross-correlation functions for the families of spreading sequences; dotted line – sequences defined by original Golay-Hadamard matrix, solid line - modified quadri-phase sequences

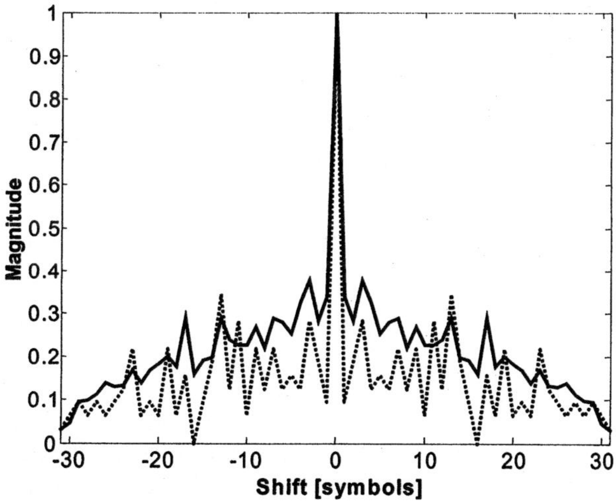

Figure 6. Peak magnitude of aperiodic auto-correlation functions for the families of spreading sequences; dotted line – sequences defined by original Golay-Hadamard matrix, solid line - modified quadri-phase sequences

4. SYSTEM SIMULATION RESULTS

To better check the usefulness of the designed sequence sets for DS CDMA applications, we have simulated a multi channel DS CDMA system utilizing BPSK signals spread by these signatures. We transmitted 500 frames of 524 bits per frame in every channel with 8 randomly selected channels simultaneously active.

The simulation has been performed for the set of 26-chip modified quadri-phase sequences, for 32-chip bipolar modified sequences, and for the set of 32-chip quadri-phase sequences. The achieved average bit error rates (BER) are 0.0030, and 0.0011, respectively, and the histograms of errors distributions are given in Fig. 7, 8, and 9.

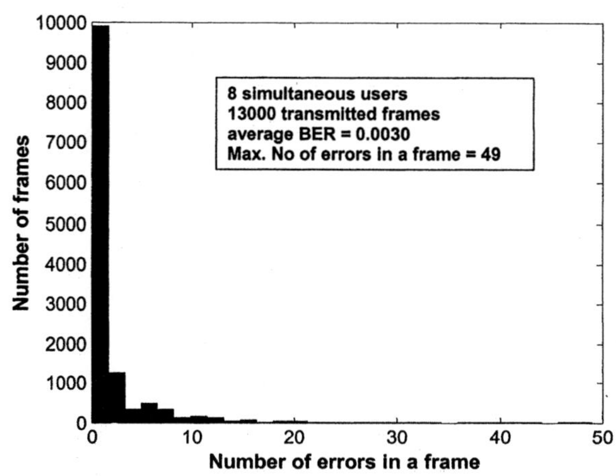

Figure 7. Histogram of a number of errors in transmitted frames for the DS CDMA system utilizing modified quadri-phase Golay-Hadamard spreading sequences of length 26

5. CONCLUSIONS

In this chapter, we considered an application of complementary Gold sequences to create orthogonal spreading sequences for asynchronous DS CDMA applications. We presented the technique of creating the original Golay-Hadamard matrices, and later a method to modify their correlation characteristics to achieve better performance in case of asynchronous operation.

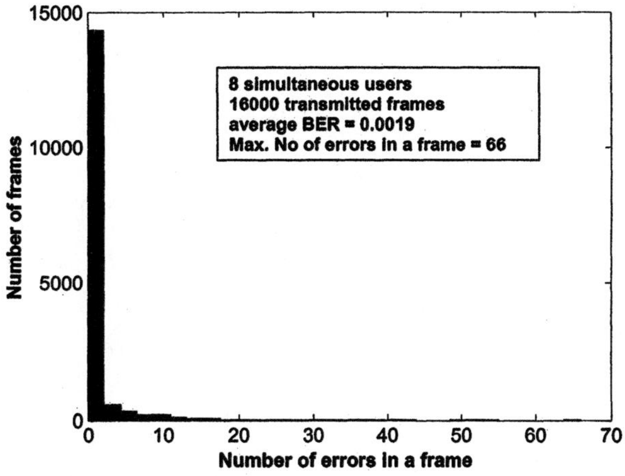

Figure 8. Histogram of a number of errors in transmitted frames for the DS CDMA system utilizing modified bipolar Golay-Hadamard spreading sequences of length 32

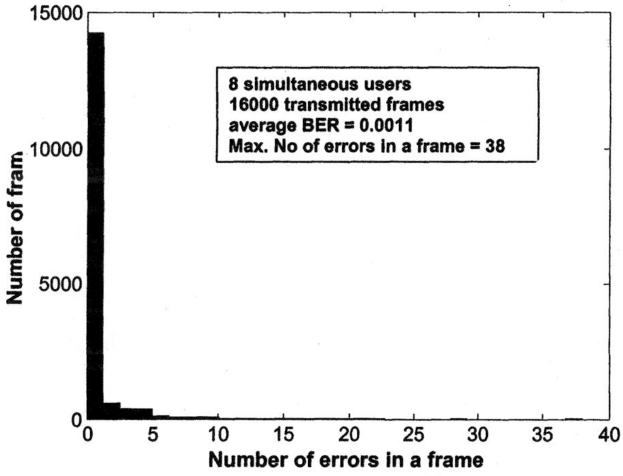

Figure 9. Histogram of a number of errors in transmitted frames for the DS CDMA system utilizing modified quadri-phase Golay-Hadamard spreading sequences of length 32

In the considered cases, i.e. bipolar and quadri-phase sequences, there are only limited numbers of modifications possible for a given sequence length. Unfortunately, examining all of them becomes impractical even for a modest sequence length, e.g. $N = 26$ or 32. In the paper, we used a random search to find the appropriate modifications. However, other, more advanced search methods may produce even better designs.

REFERENCES

[1] R.Steele, "Introduction to Digital Cellular Radio," in R.Steele and L.Hanzo (eds), *Mobile Radio Communications,* 2nd ed., IEEE Press, New York, 1999.

[2] H.F. Harmuth, *Transmission of Information by Orthogonal Functions*, Springer-Verlag, Berlin, 1970.

[3] M.B.Pursley, "Performance Evaluation for Phase-Coded Spread-Spectrum Multiple-Access Communication – Part I: System Analysis," *IEEE Trans. on Commun.*, vol. COM-25, pp. 795-799, 1977.

[4] A.Duel-Hallen, J.Holtzman, and Z.Zvonar, "Multiuser Detection for CDMA Systems," *IEEE Personal Communications*, April 1995, pp. 46-58.

[5] 3GPP, Third Generation Project Partnership, www.3gpp.org (accessed 20/11/01)

[6] B.J.Wysocki and T.Wysocki, "Optimization of Orthogonal Polyphase Spreading Sequences for Wireless Data Applications", *Vehicular Technology Conference,2001. VTC 2001 Fall*, Vol.: 3, 2001 pp. 1894 –1898.

[7] P.Z.Fan, M.Darnell and B.Honary, "Orthogonal complementary pairs of sequences," 2nd *International Symposium on Communication Theory and Applications*, Ambleside, U.K. July 1993, pp.118-121.

[8] B.Guangguo, "Methods of constructing orthogonal complementary pairs and sets of sequences," *IEEE ICC'85,* Chicago, USA, pp.839-843.

[9] T.Weng and B.Guangguo, "Application of orthogonal complementary pair of sequences to CDMA-QAM communication system," *Int. Conf. On Communications Technology, ICCT'87*, Nanjing, China, Nov. 1987, pp.826-829.

[10] P.Fan and M.Darnell, *Sequence Design for Communications Applications,* John Wiley & Sons, New York, 1996.

[11] J. Seberry and M.Yamada, "Hadamard matrices, sequences, and block designs," in J.H.Dinitz and D.R.Stinson, eds.: *Contemporary Design Theory: A Collection of Surveys,* John Wiley & Sons, Inc., pp.431-437, 1992.

[12] W.D.Wallis, A.Penfold Street and J.Seberry Wallis, "Combinatorics: Room Squares, Sum-free Sets, Hadamard Matrices," *Lecture Notes in Mathematics*, 292, Springer Verlag, Berlin-Heidelberg-New York, 1972.

[13] P.Delsarte, J.-M.Goethals, J.J.Seidel, Spherical codes and designs, *Geom. Dedicata* 6, 1977, 363-388.

[14] R.Craigen, W.Holzman and H.Kharaghani, "Complex Golay Sequences: Structure and Applications", Report, 2001.

[15] R.Craigen, "Complex Golay sequences," *J. Combin Math. and Comb. Comput.,* 15, pp.1251-1253, 1994.

[16] W.H.Holzmann and H.Kharaghani, "A computer search for complex Golay sequences," *Australasian Journal of Combinatorics*, No 10, pp.251-258, 1994.

[17] I.Oppermann and B.S.Vucetic, "Complex spreading sequences with a wide range of correlation properties," *IEEE Trans. on Commun.*, vol. COM-45, pp.365-375, 1997.

Chapter 15

PUM-BASED TURBO CODES

L Fagoonee*, Prof B Honary*, C Williams**
**Dept. of Communication Systems, Lancaster University, Lancaster, UK*

***Qinetiq, Malvern, UK*

Abstract: Partial Unit Memory (PUM) codes have excellent distance properties and we show how to construct recursive code matrices for application into turbo schemes. An iterative max-log-MAP decoding scheme is proposed for the resulting trellises with multi-input labels.

Key words: Partial Unit Memory Codes, Turbo Codes, Maximum A Posteriori (MAP) algorithm

1. INTRODUCTION

Lauer [1] introduced Partial Unit Memory (PUM) codes in 1979. PUM codes, which can be described as multiple-input convolutional codes, are optimal in the sense of having maximum free distance for a given code rate, number of encoder inputs and number of encoder memory cells. Their advantage over standard convolutional codes is the reduced number of states in their trellis for the same number of encoder inputs because only a fraction of these are shifted to the memory cells. On the other hand, PUM codes can attain larger free distances than block codes of the same codeword size and rate. Moreover, the block structure of the code facilitates synchronization. Another advantage of PUM codes is that their block length is such that they can be chosen to agree with the byte or word length of the target microprocessor, allowing further simplification during implementation. A number of non-systematic generator matrices have been searched and published whose distance properties sometimes exceed equivalent convolutional codes of the same block size and memory complexity [1,2]. In recent years turbo codes [4] have proved themselves when used in conjunction with iterative decoding schemes. However, to maintain a reasonably high throughput, turbo codes use systematic constituent codes and transmit the information bits only once together with a number of sets of parity bits. We have previously attempted to construct systematic PUM codes [3], but did not allow feedback in the encoder structure thereby limiting the free distance of the resulting code and hence their error correction ability when used together with maximum-likelihood decoding. In this paper, we describe the construction of minimal recursive systematic PUM generator encoder structures from equivalent non-systematic generator matrices, which generate codewords whose free distances reach or are very close to the upper bound defined in [1].

This chapter is organized as follows. In section 2, PUM codes and their properties are described. In section 3, we show, by means of an example, how to construct equivalent systematic PUM generator matrices from known non-systematic generator matrices. The resulting encoder structure can sometimes be minimized to an equivalent structure with fewer delay elements. This minimization procedure is described in section 4 by continuing the example from the previous section. The max-log MAP decoding algorithm for PUM component decoders in a turbo scheme is described in section 5. In section 6, results from the simulation of a turbo code with two constituent recursive systematic PUM (RSPUM) codes in a Gaussian channel are illustrated. This turbo code is compared with an equivalent recursive systematic convolutional (RSC) based turbo code. By equivalent, we mean that these component codes have the same free distance, constraint length and code rate.

Simulation results indicate that the RSPUM turbo code outperforms the equivalent RSC turbo code. In general, the performance of RSPUM turbo codes is comparable or in many cases better than that of the equivalent classical RSC turbo code.

2. PUM TURBO CODES

2.1 Code Properties

Structurally PUM codes occupy an intermediate position between block and convolutional codes. Block codes are not affected by previous memory whereas the outputs of a convolutional encoder, which satisfies gcd(n, k)=1, depend on the current input word, usually 1 bit long, and more than one previous input words.

A codeword, $\boldsymbol{x}_t$, of the (n, k, μ, d_{free}) PUM code is a function of the current input word of k information bits, $\boldsymbol{u}_t$ and a fraction μ of the previous input word, $\boldsymbol{u}_{t-1}$. This is expressed by equation 1.

$$\boldsymbol{c}_t = [\boldsymbol{u}_t\ \boldsymbol{u}_{t-1}]\mathbf{G}(\mathrm{D}) = \boldsymbol{u}_t\mathbf{G}(0) + \boldsymbol{u}_{t-1}\mathbf{G}(1) \qquad (1)$$

G(D) is the generator or encoding matrix of the code as a function of the delay D, which in this case will be the equivalent to t. For PUM codes, G(D) is non-zero only when D is equal to 0 and 1. **G(0)** and **G(1)** are generator matrices of size $k \times n$, where n is the codeword length. The rank of **G(1)** is μ, where $\mu < k$. μ determines the state complexity of the state diagram and trellis of the code. The addition and multiplication operations are modulo-2 for binary codes. The free distance, d_{free} is the minimum weight of a codeword that diverges from the all-zero path in the encoder standard trellis and returns to the all-zero path for the first time.

A standard trellis can represent a PUM code trellis with 2^{μ} starting and ending states and 2^k branches leaving and entering each node, respectively. Since a PUM code is characterized by $k>\mu$, there are $2^{k-\mu}$ parallel branches between any two states in the trellis. Each branch is labeled with n symbols. For codes encoded by non-systematic generator matrices, the trellis labels will all be parity bits. In the case of systematic code structures, the trellis labels will each be made of k information bits and n-k parity bits.

For example, the standard trellis of a (4,2,1) PUM code, systematic or non-systematic, is shown in Fig. 1.

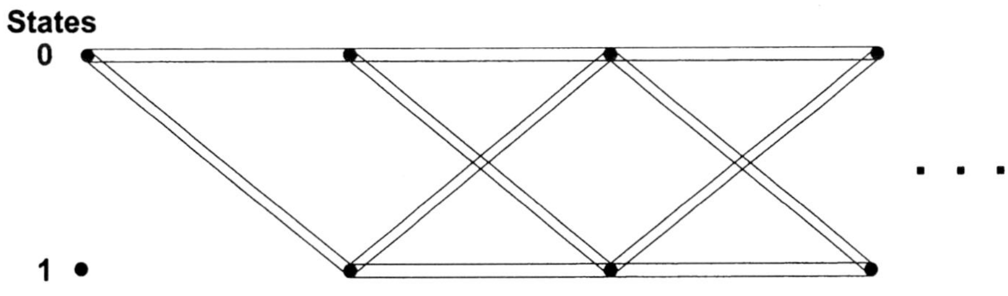

Figure 1. Standard trellis of an (4,2,1) PUM code

2.2 Encoder Structure

The typical encoder structure for a PUM code, illustrating how the number of states in a trellis is always less than a convolutional code with the same constraint length, is shown by example of a simple code.

A generator matrix for the (4,2,1,4) PUM code is shown in equation 2. It is worthwhile noting that the maximum attainable free distance [1] for a (4,2,1) PUM code with constraint length 2 is 4.

$$G(D) = \begin{bmatrix} 1 & 1 & 1 & 1 \\ 1+D & 0 & 1 & D \end{bmatrix} \tag{2}$$

Equation 3 shows the equivalent **G**(0) and **G**(1).

$$G(0) = \begin{bmatrix} 1 & 1 & 1 & 1 \\ 1 & 0 & 1 & 0 \end{bmatrix}, \; G(1) = \begin{bmatrix} 0 & 0 & 0 & 0 \\ 1 & 0 & 0 & 1 \end{bmatrix} \tag{3}$$

The encoder structure, in terms of shift registers, for the above generator matrix is shown in Fig. 2. The first row of G(1), consisting of all zeros, is represented by the highlighted unused bit from the previous input u_{t-1}. On the other hand, the first row of G(0) with all ones is indicated by a contribution of the first bit of the current input u_t towards the 4-bit codeword.

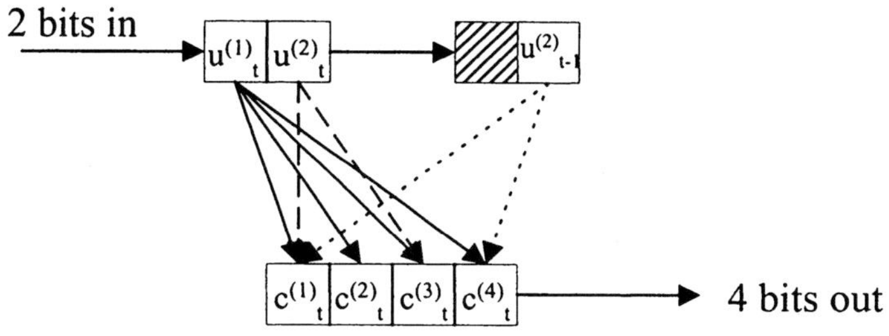

Figure 2. Encoder for non-systematic (4,2,1,4) PUM code

The above encoder structure is non-systematic as all the information bits are encoded.

3. CONSTRUCTION OF RSPUM CODE STRUCTURES

3.1 Procedure

Systematic convolutional-type generator matrices are in general simpler to implement than non-systematic generator matrices. The former have trivial right inverses, but unless rational generator matrices are used, i.e. feedback is allowed in the encoder, they are in general less powerful when used together with maximum-likelihood decoding. Since a systematic generator matrix has a sub-matrix that is a $k \times k$ identity matrix, we immediately have the following theorem: A systematic generator matrix is a systematic encoding matrix.

This approach solves the problem of free distance limitation on systematic codes, by allowing feedback in the generator matrix.

Every convolutional matrix is equivalent to a systematic rational encoding matrix. If an encoding matrix **G**(D) has some $k \times k$ sub-matrix T(D) whose determinant is a delay free polynomial, pre-multiplication by the inverse of such a matrix yields an equivalent systematic encoding matrix [5].

$$G_{sys}(D) = T(D)^{-1} G(D) \tag{4}$$

However, if the determinant of the leftmost $k \times k$ sub matrix of **G**(D) is not a delay free polynomial, then it is always possible, by permuting the columns of **G**(D), to find a generator matrix **G**'(D) whose leftmost $k \times k$ sub matrix has a determinant which is a delay free polynomial, where **G**'(D) encodes an equivalent code.

Hence, without loss of generality, a systematic encoding matrix can be written as in equation 5.

$$G(D) = \begin{bmatrix} I_k & R(D) \end{bmatrix} \tag{5}$$

3.2 Example

The construction steps are best illustrated by means of an example. The equivalent systematic matrix of the aforementioned (4,2,1,4) PUM code will be derived from its non-systematic matrix given in equation 2.

3.2.1 Step 1

The first two rows of G(D) are chosen to form T(D), as shown in equation 6.

$$T(D) = \begin{bmatrix} 1 & 1 \\ 1+D & 0 \end{bmatrix} \tag{6}$$

The inverse of T(D) is a delay-free polynomial as shown in equation 7.

$$T(D)^{-1} = \frac{1}{1+D}\begin{bmatrix} 0 & 1 \\ 1+D & 1 \end{bmatrix} \tag{7}$$

The systematic generator matrix $G_{sys}(D)$ is obtained as in equation 4.

$$G_{sys}(D) = \begin{bmatrix} 1 & 0 & \frac{1}{1+D} & \frac{D}{1+D} \\ 0 & 1 & \frac{D}{1+D} & \frac{1}{1+D} \end{bmatrix} \tag{8}$$

3.2.2 Step 2

The controller canonical form of $G_{sys}(D)$, obtained directly from equation 8, contains 2 storage registers as shown in Fig. 3.

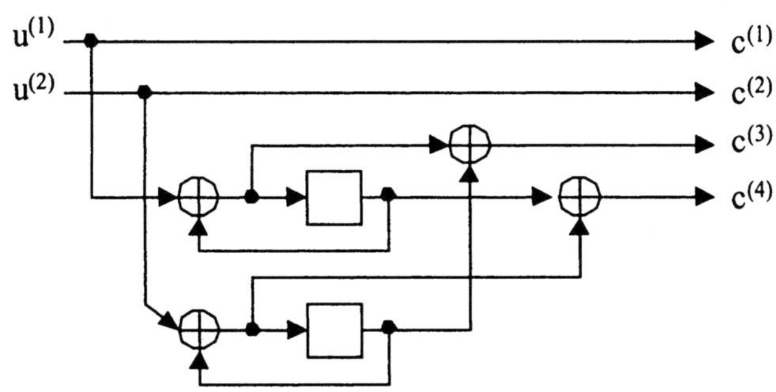

Figure 3. Systematic encoder structure of (4,2,1,4) PUM Code

However, a minimal encoder with only 1 register can be realized. This minimization process [5] is described in the next section.

4. LOW COMPLEXITY RSPUM ENCODER STRUCTURE

Minimization is carried out by analyzing the redundancy, i.e. any identical outputs for the same inputs, for varying present states.

Let $s^{(1)}_t$ and $s^{(2)}_t$ be the states of the first and second delay elements at time t, $u^{(1)}_t$ and $u^{(2)}_t$ be the two inputs to the encoder at time t, $c^{(3)}_t$ and $c^{(4)}_t$ be the two parity bits calculated by the encoder at time t. The two systematic outputs can be ignored for this process, as they remain unchanged.

The next states of the delay elements at time t+1, $s^{(1)}_{t+1}$ and $s^{(2)}_{t+1}$, and the outputs of encoder at time t can be written as a function of the states and the inputs at time t.

$$s^{(1)}_{t+1} = s^{(1)}_t + u^{(1)}_t \tag{9}$$

$$s^{(2)}_{t+1} = s^{(2)}_t + u^{(2)}_t \tag{10}$$

$$c^{(3)}_t = s^{(1)}_{t+1} + s^{(2)}_t = s^{(1)}_t + s^{(2)}_t + u^{(1)}_t \tag{11}$$

$$c^{(4)}_t = s^{(1)}_t + s^{(2)}_{t+1} = s^{(1)}_t + s^{(2)}_t + u^{(2)}_t \tag{12}$$

These equations are used to build Table 1, which represents the successor states and outputs ($s^{(1)}_{t+1}\, s^{(2)}_{t+1}/c^{(3)}_t\, c^{(4)}_t$) as a function of the present state ($s^{(1)}_t\, s^{(2)}_t$) and the inputs ($u^{(1)}_t\, u^{(2)}_t$).

Table 1. All successor states/outputs combinations

Present State $s^{(1)}_t\, s^{(2)}_t$	Inputs $u^{(1)}_t\, u^{(2)}_t$			
	00	01	10	11
	$s^{(1)}_{t+1}\, s^{(2)}_{t+1}/c^{(3)}_t\, c^{(4)}_t$			
00	**00/00**	**01/01**	**10/10**	**11/11**
01	**01/11**	**00/10**	**11/01**	**10/00**
10	**10/11**	**11/10**	**00/01**	**01/00**
11	**11/00**	**10/01**	**01/10**	**00/11**

In order to find which present states are equivalent, we start by merging those states that correspond to the same outputs. Hence, we obtain partition P1: {00,11}, {01,10}.

The states that bel(

The next stage is to determine whether this set can be further broken down into a partition P_2 whose states are 2-equivalent. Two states are 2-equivalent if they are 1-equivalent and their successor states are 1-equivalent. In this case, $P_2=P_1$. In general, partitioning is repeated until $P_{i+1}=P_i$.

In order to obtain a linear realization of the minimal encoder, we let 0 represent the state {00,11} and 1 represent {01,10}. This is shown in Table 2 which represents the successor state and outputs ($s_{t+1}/c^{(3)}_t\ c^{(4)}_t$) as a function of the present state (s_t) and the inputs ($u^{(1)}_t\ u^{(2)}_t$).

Table 2. Minimized successor state/output combinations

Present State s_t	Inputs $u^{(1)}_t\ u^{(2)}_t$			
	00	01	10	11
	$s_{t+1}/c^{(3)}_t\ c^{(4)}_t$			
0	**0/00**	**1/01**	**1/10**	**0/11**
1	**1/11**	**0/10**	**0/01**	**1/00**

The last step before construction of the minimized encoder structure is to re-write the successor state and outputs as a function of the inputs and present state. The successor state can be isolated in Table 2 to yield Table 3, to define its relationship with the present state and inputs, shown in equation 13.

Table 3. Successor state

Present State s_t	Inputs			
	00	01	10	11
	s_{t+1}			
0	**0**	**1**	**1**	**0**
1	**1**	**0**	**0**	**1**

$$s_{t+1} = s_t + u^{(1)}_t + u^{(2)}_t \tag{13}$$

Similarly, Tables 4 and 5 show the relationship between each output respectively with the present state and inputs. Equations 14 and 15, in turn, indicate the relationship algebraically.

Table 4. First coded bit

Present State (s_t)	Inputs ($u^{(1)}_t\ u^{(2)}_t$)			
	00	01	10	11
	$c^{(3)}_t$			
0	**0**	**0**	**1**	**1**
1	**1**	**1**	**0**	**0**

$$c^{(3)}_t = s_t + u^{(1)}_t \tag{14}$$

Table 5. Second coded bit

Present State s_t	Inputs ($u^{(1)}_t\ u^{(2)}_t$)			
	00	01	10	11
	$c^{(4)}_t$			
0	**0**	**1**	**0**	**1**
1	**1**	**0**	**1**	**0**

$$c^{(4)}_t = s_t + u^{(2)}_t \tag{15}$$

The minimized encoder structure is constructed from equations 13 to 15, and shown in Fig. 4.

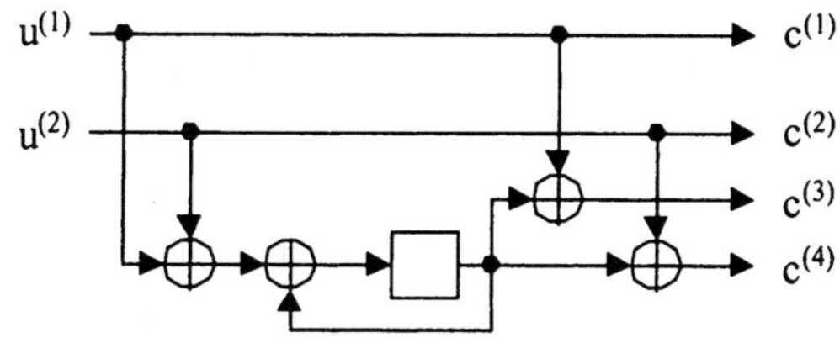

Figure 4. Systematic minimized encoder structure for (4,2,1,4) PUM code

5. MULTI-STAGE DECODING FOR TURBO CODES

Fig. 5 shows a section of the decoding trellis, including branch labels, of the realization shown in Fig. 4. The labels are written in the order $c^{(1)}{}_t c^{(2)}{}_t c^{(3)}{}_t c^{(4)}{}_t$. The edge complexity is only 4 branches per information bit (bpi).

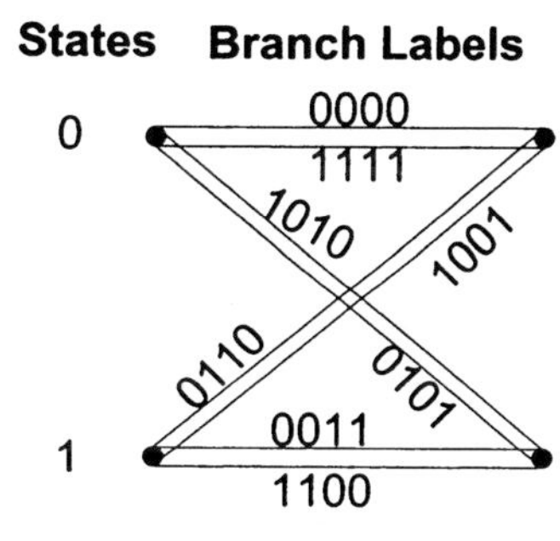

Figure 5. Trellis section of systematic (4,2,1,4) PUM code

The recursive systematic PUM codes are then incorporated as component codes in a parallel-concatenated scheme, also called turbo scheme [4]. Iterative decoding between the component decoders is achieved in the same way as that used by traditional turbo codes based on recursive systematic convolutional codes. However, the max-log MAP decoding algorithm [6] for each component code is slightly modified to allow for multi-input trellis labels.

The multi-input max-log MAP decoding algorithm is summarized by equations 16 to 20. The forward and backward recursion node metrics, α and

β are a function of the branch transition probability Γ for all possible K inputs allowing transition from s' to s.

$$\log \Gamma_t(s', s) = \frac{1}{2}\left[\sum_{k=0}^{K-1} u_{t,k} L(u_{t,k}) + \sum_{n=0}^{N-1} x_{t,n} L_c y_{t,n}\right] \tag{16}$$

$$\log \alpha_t(s) = \max_{s'}\left(\log \Gamma_t(s', s) + \log \alpha_{t-1}(s')\right) \tag{17}$$

$$\log \beta_{t-1}(s') = \max_{s}\left(\log \Gamma_t(s', s) + \log \beta_t(s)\right) \tag{18}$$

The a posteriori LLR for each of the k symbols at time t is the sum of the *a priori* LLR, the LLR of the channel-compensated received value and the extrinsic LLR of the symbol in question.

$$L(\hat{u}_{t,j}) = L(u_{t,j}) + L_c y_{t,j} + \log \gamma_{t,j}(s', s) \tag{19}$$

The extrinsic transition likelihood for symbol j at time t, $\gamma_{t,j}$ is given by equation 20.

$$\log \gamma_{t,j}(s', s) = \frac{1}{2}\left[\sum_{\substack{k, \\ k \neq j}} u_{t,k} L(u_{t,k}) + \sum_{\substack{n, \\ n \neq j}} x_{t,n} L_c y_{t,n}\right] \tag{20}$$

6. SIMULATION RESULTS

A parallel-concatenated code with two component (4,2,1,4) RSPUM codes is constructed to yield a rate $^1/_3$ turbo code. Fig. 6 shows its performance with varying iterations in an AWGN channel for an interleaver size of 10000. Fig. 7 illustrates that the RSPUM turbo code performance is no different to the traditional RSC turbo codes with varying interleaver sizes.

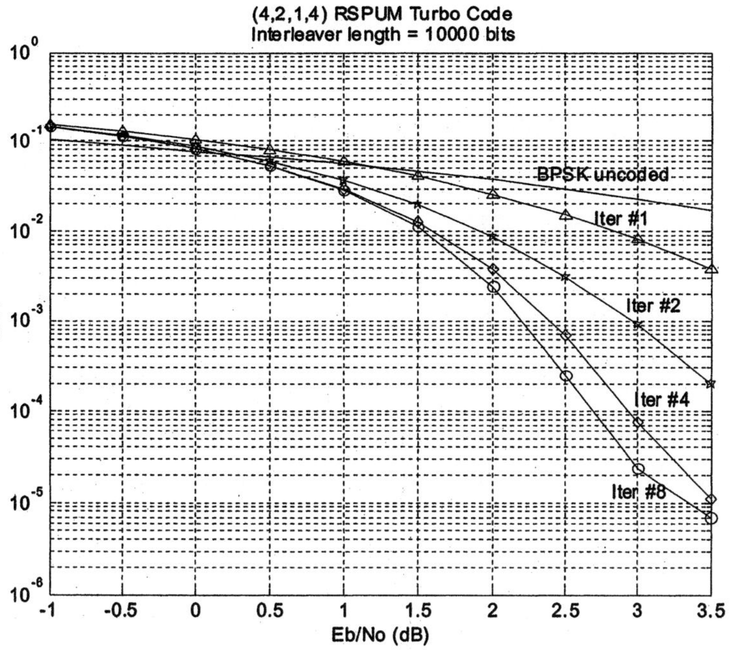

Figure 6. Performance of (4,2) RSPUM turbo code for up to 8 iterations

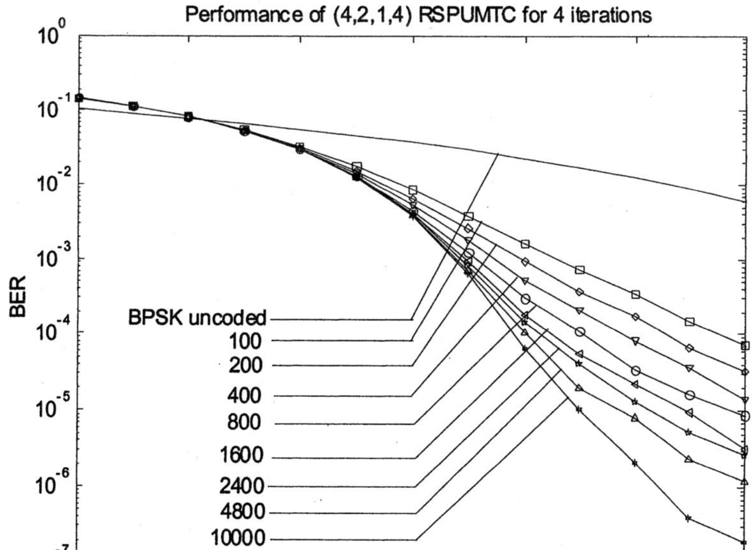

Figure 7. Performance with varying interleaver size

The encoder structure of the equivalent rate ½ recursive systematic convolutional code with constraint length 2 is shown in Fig. 8. Its trellis has 2 states and a branch complexity of 4 bpi. The code has a free distance of 4 over 2 information symbols.

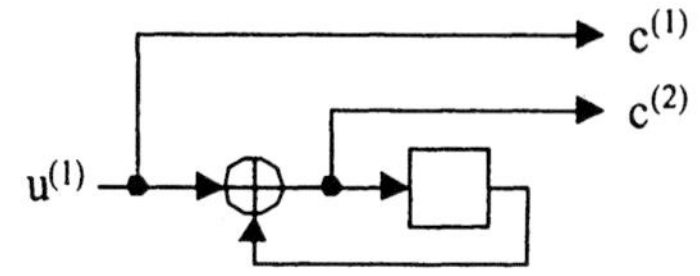

Figure 8. Equivalent RSC encoder structure

The relative performance of the rate $^1/_3$ turbo decoders with two constituent RSC codes and RSPUM codes respectively is illustrated in Fig. 9. Both turbo encoders use an interleaver size of 10000.

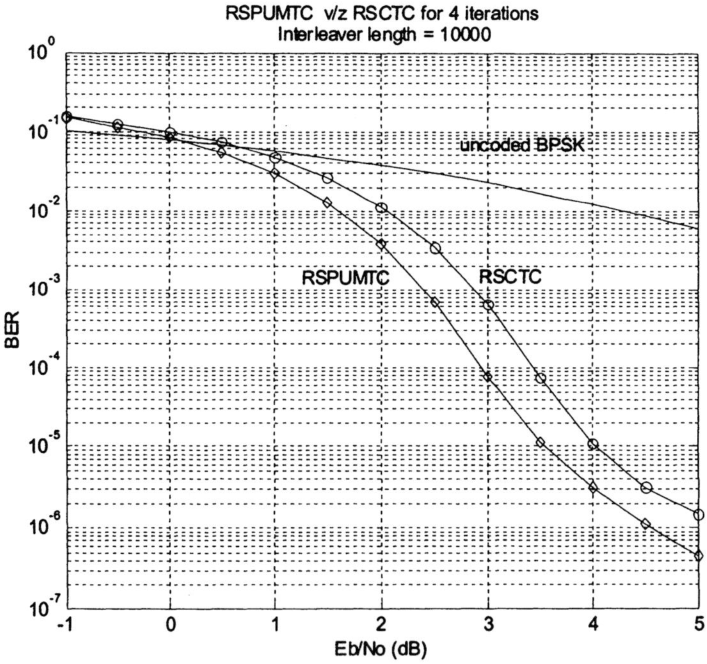

Figure 9. Relative performance of equivalent RSPUM and RSC turbo codes

Overall, the RSPUM turbo code outperforms the equivalent RSC turbo code by about 0.5 dB.

7. CONCLUSION

In this paper, the example of the simple (4,2,1,4) PUM code was used to demonstrate construction of the low complexity and maximum distance recursive systematic encoder structure. Hence recursive systematic PUM code structures can be efficiently constructed from known non-systematic generator matrices in order to retain their good distance properties. These encoder structures can be further minimized to yield encoder structures with fewer memory elements. The resulting code trellis has a very low decoding complexity and can be used effectively in turbo code systems. Simulation results showed that the performance of the (4,2,1,4) RSPUM as a constituent code in a rate $^1/_3$ turbo code system outperforms the equivalent RSC turbo code.

REFERENCES

[1] G S Lauer, "Some Optimal Partial Unit-Memory Codes," *IEEE Trans. Inf. Theory*, Vol. IT-25, No. 2, pp. 240-243, March 1979.

[2] L N Lee, "Short, unit-memory, byte-oriented, binary convolutional codes having maximal free distance," *IEEE Trans. Inf. Theory*, vol. IT-22, pp. 349-352, May 1976

[3] L Fagoonee, B Honary and C Williams. "Error Protection Using PUM Turbo Codes for Future Mobile Communication Systems," *2nd Int. Conf. on 3G Mobile Communication Technologies*, London, March 2001.

[4] C Berrou, A Glavieux and P Thitimajshima, "Near Shannon Limit Error Correcting Coding and Decoding: Turbo Codes (1)," *Proc. IEEE Int. Conf. Commun.*, Geneva, Switzerland, pp. 1740-1745, May 1993.

[5] R Johannesson and K S Zigangirov, *Fundamentals of Convolutional Coding*, IEEE Press, 1999.

[6] J Hagenauer, " Iterative Decoding of Binary Block and Convolutional Codes," *IEEE Trans. Info. Theory*, Vol. 42, No. 2, 1996.

ACKNOWLEDGEMENT

This work is funded under the Ministry of Defense Applied Research Program. The authors would also like to thank Prof. Shavgulidze for his suggestions.

Chapter 16

A CODE FOR SEQUENTIAL TRAITOR TRACING

Reihaneh Safavi-Naini and Yejing Wang
School of Information Technology and Computer Sciene, University of Wollongong, Wollongong 2522, Australia

Abstract: In a pay-TV system a subscriber uses a decoder to decrypt the broadcasted signal. Each decoder contains a unique set of decryption keys that is used to identify the owner and allow him to decrypt the data aimed at him. Traitor tracing schemes ensure that at least one of the at most c colluders, who construct a pirate decoder[1] to illegally access the data, can be identified. Sequential tracing schemes provide protection against colluders who *re-broadcast* the decrypted content to make it available for un-authorized users. A sequential tracing scheme ensures that all colluders are identified and disconnected from the system. The main construction of sequential tracing schemes is from error-correcting codes that satisfy a bound on their minimum distance. We show that the known codes that satisfy this bound require large alphabets and so are demanding on the underlying watermarking codes. We give the construction of a new error-correcting code that satisfies the required bound and protects against c colluders using the smallest possible (asymptotically) alphabet size.

Key words: traceability, sequential traitor tracing, error-correcting codes, trace function.

1. INTRODUCTION

Protecting digital content against illegal copying and redistribution is one of the key problems facing the owners and distributors of digital content.

[1] This research is in part supported by Australian Council Grant Number 227 26 1008.

Access control methods such as encryption schemes ensure that a digital object is only accessible to the person who owns the key information and so has paid the required fees. However when the content is decrypted, a malicious buyer can make illegal copies of the object and re-distribute it.

Traitor tracing schemes provide protection against illegal copying and redistribution of digital objects. A possible example application of traitor tracing is pay-TV systems. The TV station encrypts the content and broadcasts it to all users. Each user has a decoder box with a set of keys that allows him/her to decrypt the content. However, some subscribers may collude to produce a pirate decoder by using some of their keys. A traitor tracing scheme allows a broadcaster to identify at least one colluder by testing the keys inside a pirate decoder. Traitor tracing schemes were first introduced by Chor et al [3], and later studied by various authors [9,14,15,7,2,4,12,17].

Dynamic traitor tracing was introduced by Fiat and Tassa [6] to provide protection against *re-broadcasting* of the content after it is decoded. Safavi-Naini and Wang [11] showed that dynamic traitor tracing schemes are vulnerable to the delayed rebroadcast attack where the colluders rebroadcast the content after a delay. They proposed a new type of system called *sequential traitor tracing*, that is secure against this attack and gave a construction from error correcting codes whose minimum Hamming distance satisfy a particular lower bound. It was shown that Reed-Solomon codes and algebraic geometry codes satisfy this bound and so can be used to construct sequential tracing schemes.

In this paper we construct a new error-correcting code with this property, and show that compared to Reed-Solomon codes and algebraic geometry codes, it requires a smaller size alphabet while maintaining the same level of collusion security, hence resulting in evaluate performance of more efficient tracing scheme. The paper is organised as follows. In section 2 we review sequential traitor tracing schemes. Section 3 gives our construction. In Section 4 we evaluate performance of the new construction.

2. MODEL AND DEFINITIONS

In sequential traitor tracing schemes the *protected content* is divided into L *segments* and a watermarking code $W = \{1,2,\cdots,q\}$ is used to mark each segment to produce q *versions* of that segment. Let $U = \{u_1,u_2,\cdots,u_N\}$ denote the set of users. User u_i receives a sequence of marks that the broadcaster allocates to him according to a *mark allocation table* M. The table has N rows and L columns and $M(i, j)$ is an element of W and is the mark allocated to user u_i in segment j. There is a *feedback sequence* $F = ()$ which is initialised to the empty sequence. In segment j, the content provider sends the kth version to all users for whom $M(i, j)$ is k and observes the feedback. The feedback signal $f_j \in W$ is appended to F_{j-1} to construct $F_j = (f_1, f_2, \cdots, f_j)$ which is used to identify traitors. A feedback sequence F is called a c*-feedback sequence* if it can be generated by a colluder set of size at most c. A traitor is traced by examining the feedback sequence. When a traitor is found he is disconnected and tracing continuous for the next traitor until all traitors are found. After d segments all traitors are found and the tracing algorithm *converges*.

Definition An (M, A) sequential c-traceability scheme consists of a mark allocation table M and a tracing algorithm A,

1. $M = [m_{ij}]$ is an $N \times L$ array with entries from W;
2. A is a mapping $A: W^* \to 2^U$ with the property that for any c-feedback sequence F, there exists a sequence of integers $0 < d_1 < d_2 < \cdots < d_k = d \le L, k \le c$, such that

$$A(F_j) = \begin{cases} T_j, & \Phi \ne T_j \subseteq U, j = d_1, d_2, \cdots d_k \\ \Phi, & otherwise \end{cases} \quad \text{and} \quad T_1 \cup T_2 \cup \cdots \cup T_k = C.$$

In the above definition, d is the *convergence* length of tracing. That is d is the maximum number of steps in the algorithm to trace all (up to c) traitors. It is shown [11] that a sequential c-traceability scheme can be constructed from a q-ary error-correcting codes. We denote by $(L,N,D)_q$-ECC an error-correcting code over a q-ary alphabet with length L, minimum Hamming distance D and having N codewords. A q-ary linear error-correcting code of length L and of dimension k is denoted by $[L,k,D]_q$.

Theorem 2.1 ([11]) Suppose there is an $(L,N,D)_q$-ECC, Γ. If

$$D > (1 - \frac{1}{c^2})L + \frac{1}{c}, \tag{1}$$

then Γ is a sequential c-traceability scheme which converges in $d=c(L-D+1)$ steps.

Theorem 2.1 shows that sequential c-traceability schemes can be constructed from error-correcting codes with large Hamming distances. From (1) we have that the maximum collusion size c satisfies

$$c < \frac{-1+\sqrt{1+4L(L-D)}}{2(L-D)}.$$

Examples of error-correcting codes that satisfy (1) are Reed-Solomon codes (RS-codes) and algebraic geometry codes (AG-codes). An $[L,k,N]_q$ RS-code is a linear q-ary code with L=q-1, D=L-k+1 and $N=q^k$. For a RS-code we have

$$c < -\frac{1}{2(k-1)} + \sqrt{\frac{1}{4(k-1)^2} + \frac{q-1}{k-1}}.$$

For these codes the number of codewords grows with increasing k or q. However according to the above equation an increase in k results in a reduction of c and so to fix c and increase the number of codewords, q must be increased. The drawbacks of increasing q will be discussed in Section 4.

An AG-code is a $[L,k,L+1-k-g]_q$ code. It is known [16] that an AG-code $[L,k,L+1-k-g]_q$ exists if there exists an algebraic curve of genus g over GF(q) having L rational points. Using the results in [16] implies that an AG-code $[L,k,L-k-1]_q$ exists for any k, 1≤k≤L-2, and L is at most $q+4\sqrt{q}+1$ for g≤2. Using AG-codes for sequential tracing has a similar drawback. The following Theorem gives a class of codes that overcome the drawback of RS-codes and AG-codes.

Theorem 2.2 ([1],[2]) For any positive integers p, N, let L=8plog N. Then there exists a $(L,N,D)_{2p}$-ECC where D>(1-1/p)L.

For sequential traceability scheme obtained from this code, we have

$$c < -\frac{p}{2L} + \sqrt{\frac{p^2}{4L^2} + p}$$

which for large N (and so large L) is approximately equal to $\sqrt{p}$, or $\sqrt{q/2}$. The theorem only shows existence of such a code but is non-constructive. We will construct a new code in the next section, that overcomes the drawback of the RS-codes and AG-codes, but the traceability c is not as good as the one given in Theorem 2.2.

3. THE NEW CONSTRUCTION

Let GF(q) and GF(q^k) be fields of q and q^k elements, respectively. The trace function $Tr_{qk|q}$: GF(q^k)→GF(q) of GF(q^k) over GF(q) is given by,

$$Tr_{q^k|q}(x) = x + x^q + x^{q^2} + \cdots + x^{q^{k-1}}.$$

When it is clear from the context we will use Tr(x) instead of $Tr_{qk|q}(x)$. The following proposition can be found in [5].

Proposition 3.1 For any a∈GF(q),

$$|\{x \in GF(q^k):Tr(x)=a\}|=q^{k-1}.$$

Our construction starts with a basic construction in section 3.1 and it is then extended in section 3.2.

3.1 Basic construction

Let k and s be positive integers such that

$$k=2t,\ s<q^{k/2}+1, \text{ and } \exists r \text{ such that } r|t,\ q^r=-1 (\text{mod } s) \quad (2)$$

Note that if k and s satisfy (2) and we have $s|q^r+1$, then since r|t, we have $r|q^t+1$ and so $s|q^k-1$, that is $s|(q^t+1)(q^t-1)$.

Let $x_1,x_2,\ldots,x_{qk}$ be distinct elements of GF(q^k), and $\beta_1,\beta_2,\ldots,\beta_q \in GF(q^k)$ be q elements of GF(q^k) whose trace values are pairwise distinct. Consider the following vector

$$(Tr(\alpha x_1^s+\beta),Tr(\alpha x_2^s+\beta),\ldots,Tr(\alpha x_{qk}^s+\beta)) \quad (3)$$

where $\alpha \in GF(q^k)^*$, $\beta \in \{\beta_1,\beta_2,\ldots,\beta_q\}$, and k and s satisfying the conditions in (2). For simplicity, we denote the vector (3) by (α,β,s), and

$$z(\alpha,\beta,s)=|\{x \in GF(q^k):Tr(\alpha x^s+\beta)=0\}|.$$

Wolfmann proved the following theorem.

Theorem 3.1 ([18,19]) Let k and s be integers satisfying (2) and n be such that $ns=q^k-1$. Then $z(\alpha,\beta,s)$ is given by the following formulae

1. If $\alpha^n=\sigma_1$ and $Tr(\beta)=0$, then $z(\alpha,\beta,s)=q^{k-1}-\sigma(s-1)(q-1)q^{k/2-1}$;
2. If $\alpha^n=\sigma_1$ and $Tr(\beta)\neq0$, then $z(\alpha,\beta,s)=q^{k-1}+\sigma(s-1)q^{k/2-1}$;
3. If $\alpha^n\neq\sigma_1$ and $Tr(\beta)=0$, then $z(\alpha,\beta,s)=q^{k-1}+\sigma(q-1)q^{k/2-1}$;
4. If $\alpha^n\neq\sigma_1$ and $Tr(\beta)\neq0$, then $z(\alpha,\beta,s)=q^{k-1}-\sigma q^{k/2-1}$;

where $\sigma=(-1)^{t/2}$ and $\sigma_1=\sigma^u$ with $us=q^r+1$.

Using Theorem 3.1 the following Lemma can be proved for vectors of the form (3).

Lemma 3.1 For a fixed integer s satisfying (2) and $\alpha_1,\alpha_2\in GF(q^k)^*$, if $\alpha_1\neq\alpha_2$, then $(\alpha_1,\beta,s)\neq(\alpha_2,\beta',s)$ for all $\beta,\beta'\in\{\beta_1,\beta_2,\ldots,\beta_q\}$.

Proof Because of the property of trace functions, $Tr(x+y)=Tr(x)+Tr(y)$, we have

$$(\alpha_1,\beta,s)=(\alpha_2,\beta',s)$$

$$\text{if } Tr(\alpha_1x^s+\beta)=Tr(\alpha_2x^s+\beta')=0 \text{ for all } x\in GF(q^k)$$

$$\text{if } Tr((\alpha_1-\alpha_2)x^s+(\beta-\beta'))=0 \text{ for all } x\in GF(q^k)$$

If $\alpha_1-\alpha_2\neq0$ then the last equality leads to a contradiction because of Theorem 3.1 which gives $|\{x\in GF(q^k):Tr((\alpha_1-\alpha_2)x^s+(\beta-\beta'))=0\}|<q^k$.

Lemma 3.2 gives a necessary and sufficient condition for equality of vectors of the form (α,β,s).

Lemma 3.2 For integer s satisfying (2) and $\alpha\in GF(q^k)^*$, $\beta,\beta'\in\{\beta_1,\beta_2,\ldots,\beta_q\}$, the two vectors (α,β,s) and (α,β',s) are equal if and only if $\beta=\beta'$.

Proof Follows from definition of vectors in (3).

Let $v_1=(\alpha_1,\beta,s)$ and $v_2=(\alpha_2,\beta',s)$. The distance $D(v_1,v_2)$ between v_1 and v_2 can be obtained by using Theorem 3.1 as follows,

$$D(v_1,v_2)=|\{x\in GF(q^k):Tr((\alpha_1-\alpha_2)x^s+(\beta-\beta'))\neq 0\}|$$
$$=q^k-|\{x\in GF(q^k):Tr((\alpha_1-\alpha_2)x^s+(\beta-\beta'))=0\}|$$
$$\geq q^k-q^{k-1}-(s-1)(q-1)q^{k/2-1}$$

Let c be an integer satisfying,

$$c<\frac{-1+\sqrt{1+4q^{3k/2}(q^{k/2}+(s-1)(q-1))}}{2q^{k/2-1}(q^{k/2}+(s-1)(q-1))} \qquad (4)$$

The value on the right of (4) is the positive zero of the quadratic equation

$$(q^{k-1}+(s-1)(q-1)q^{k/2-1})x^2+x-q^k=0$$

and so if c satisfies (4) we have

$$(q^{k-1}+(s-1)(q-1)q^{k/2-1})c^2+c-q^k\leq 0$$

That is,

$$c^2(q^k-q^{k-1}-(s-1)(q-1)q^{k/2-1})\geq(c^2-1)q^k+c.$$

and so

$$D\geq(1-1/c^2)L+1/c.$$

The following theorem summarizes the above.

Theorem 3.2 For any prime power q and an even integer k, there exists an $(L,N,D)_q$-ECC in which

1. $L=q^k$, $N=q(q^k-1)$, and
2. D satisfies (1) with c given in (4).

3.2 Extension of the basic construction

To increase N, we extend the above construction to include more values of s. Let $k=2t$, and $r|t$ and $s=q^r+1$. Define

$$\Gamma = \cup_{r|t} \cup_{\alpha \in GF(q^k)^*} \{(\alpha,\beta_1,q^r+1),(\alpha,\beta_2,q^r+1),\cdots,(\alpha,\beta_q,q^r+1)\}$$

We will calculate the minimum distance D(Γ) and prove that (1) is satisfied for some c.

First we recall the concept of quadratic forms and review some results. A more complete treatment of the quadratic forms can be found in [8].

A quadratic form in k indeterminates over GF(q) is a homogeneous polynomial in $GF(q)[x_1,x_2,\ldots,x_k]$ of degree 2, or is the zero polynomial. If q is odd, a quadratic form f over GF(q) can be written as

$$f(x_1,x_2,\cdots x_k) = \sum_{i,j=1}^{k} a_{ij}x_ix_j, \text{ with } a_{ij}=a_{ji}$$

or

$$f(x_1,x_2,\cdots,x_k) = X^T AX$$

where $A=(a_{ij})_{k\times k}$ is the coefficient matrix of f, and X^T is the transpose of the column vector

$$X = \begin{pmatrix} x_1 \\ x_2 \\ \vdots \\ x_k \end{pmatrix}$$

The *rank* of a quadratic form is the rank of its coefficient matrix, and the *determinant* of a quadratic form f, denoted by det(f), is det(A). A quadratic form f in k indeterminates is called *nondegenerate* if A is of rank k. Every non-degenerate quadratic form f over GF(q), q odd, can be written as a form with a diagonal coefficient matrix

$$\begin{pmatrix} a_{11} & & & \\ & a_{22} & & \\ & & \ddots & \\ & & & a_{kk} \end{pmatrix}, \ a_{ii} \neq 0, \ \forall i. \tag{5}$$

We can consider $GF(q^k)$ as a vector space of dimension k over GF(q) and write an element $x \in GF(q^k)$ as a k-dimensional vector over GF(q). We define the trace of an element of $GF(q^k)$ as the trace of the corresponding element using the one-to-one correspondence between an element of $GF(q^k)$ and its vector space representation in $GF(q^k)$.

Lemma 3.3 (Lemma 2, [18]) Let k and r be integers such that (2r)|k. For an $\alpha \in GF(q^k)^*$, define a function $f_{\alpha,r}$ as follows

$$f_{\alpha,r} : GF(q^k) \to GF(q), f_{\alpha,r}(x) = Tr(\alpha x^{q^r+1}).$$

Then $f_{\alpha,r}$ is a quadratic form over GF(q).

It follows that

$$g = f_{\alpha_1,r_1} - f_{\alpha_2,r_2} \tag{6}$$

is also a quadratic form over GF(q).

Let **R** be the set of real numbers. *The quadratic character* of GF(q) η:$GF(q)^* \to \mathbf{R}$ is defined by $\eta(a)=1$, if a is the square of an element of $GF(q)^*$, and $\eta(a)=-1$, otherwise. We define a function v: $GF(q) \to \mathbf{R}$ by

$$v(a)=-1, \text{ if } a \neq 0, \text{ and } v(a)=q-1, \text{ if } a=0.$$

Theorem 3.3 (Theorem 6.26 and 6.27, [8]) Let q be odd, η be the quadratic character of GF(q). If f is a nondegenerate quadratic form in k indeterminates over GF(q), $\Delta=\det(f)$, then for any $a \in GF(q)$ the number of solutions of the equation $f(x_1, x_2, \cdots, x_k) = a$ in $GF(q^k)$ is given by

1. $q^{k-1}+v(a)q^{k/2-1}\eta((-1)^{k/2}\Delta)$, if k is even;
2. $q^{k-1}+q^{(k-1)/2}\eta((-1)^{(k-1)/2}a\Delta)$, if k is odd.

Now consider the code Γ. To find the distance between two codewords we count the number of components where two codewords (α_1,β,s_1) and

(α_2,β',s_2), where $s_1 = q^{r_1}+1, s_2 = q^{r_2}+1$, are different. That is we count the number of elements $x \in GF(q^k)$ such that $Tr(\alpha_1 x^{qr_1+1}+\beta)=Tr(\alpha_2 x^{qr_2+1}+\beta')$, and so we need to count the number of x such that

$$Tr(\alpha_1 x^{qr_1+1})-Tr(\alpha_2 x^{qr_2+1})=Tr(\beta')-Tr(\beta).$$

This is equal to the number of $x \in GF(q^k)$ such that $g(x)=Tr(\beta'-\beta)$, where g is a quadratic form as shown in (6).

Suppose the rank of g is m, where $m \le k$. Lemma 1 and Lemma 2 in [18] showed that $g \neq 0$, and hence $m \ge 1$. Assume that the coefficient matrix of g is given as in (5) with $a_{11},a_{22},\ldots,a_{mm}\neq 0$ and $a_{m+1,m+1}=\ldots=a_{k,k}=0$. Then for any $(x_1,\ldots,x_m,\ldots,x_k)\in GF(q^k)$ we have

$$g(x_1,\cdots,x_m,x_{m+1},\cdots,x_k)=g(x_1,\cdots,x_m,0,\cdots,0). \qquad (7)$$

Therefore we can think of g as a quadratic form in m indeterminates over GF(q). Let $a=Tr(\beta'-\beta)$. From Theorem 3.3 we know that when m is even,

$$|\{x \in GF(q^m)\colon g(x)=a\}|=q^{m-1}+v(a)^{qm/2-1}\}\eta((-1)^{m/2}\Delta),$$

and when m is odd,

$$|\{x \in GF(q^m)\colon g(x)=a\}|=q^{m-1}+q^{(m-1)/2}\eta((-1)^{(m-1)/2}a\Delta).$$

Since

$$\begin{aligned}&|\{x=(x_1,\cdots,x_m,\cdots,x_k)\in GF(q^k): g(x)=a\}|\\ &=|\{x=x_1,\cdots,x_m,0,\cdots,0)\in GF(q^k): g(x)=a\}|\,q^{k-m}\end{aligned} \qquad \text{(from (7))}$$

we obtain that

$$|\{x \in GF(q^k)\colon g(x)=a\}|=q^{k-1}+v(a)q^{k-m/2-1}\eta((-1)^{m/2}\}\Delta), \qquad (8)$$

if m is even, and

$$|\{x \in GF(q^k)\colon g(x)=a\}|=q^{k-1}+q^{k-(m+1)/2}\eta((-1)^{(m-1)/2}a\Delta), \qquad (9)$$

if m is odd.

The equalities (8) and (9) show that two codewords (α_1,β,s_1) and (α_2,β',s_2) are distinct as long as they are distinct triple. Furthermore, from (8) and (9) the distance $D(u_1,u_2)$ between $u_1=(\alpha_1,\beta,s_1)$ and $u_2=(\alpha_2,\beta',s_2)$ satisfies

$$\begin{aligned} D(u_1,u_2) &= q^k-|\{x\in GF(q^k):\ g(x)=a\}| \\ &\geq q^k-q^{k-1}-(q-1)q^{k-m/2-1} \\ &\geq q^k-q^{k-1}-(q-1)q^{k-3/2} \\ &= q^k-q^{k-1/2}-q^{k-1}+q^{k-3/2} \end{aligned}$$

So the minimum distance $D=D(\Gamma)$ satisfies

$$D\geq q^k-q^{k-1/2}-q^{k-1}+q^{k-3/2}.$$

Let c be an integer satisfying

$$c=\left\lfloor \frac{-1+\sqrt{1+4q^k(q^{k-1/2}+q^{k-1}-q^{k-3/2})}}{2(q^{k-1/2}+q^{k-1}-q^{k-3/2})} \right\rfloor. \qquad (10)$$

Then we have

$$D>(1-1/c^2)L+1/c.$$

Finally, we count the number of s. Since s is of the form $s=q^r+1$, where $r|t$, then the number of possible s is the number of divisors of t. Let $\tau(t)$ be the number of divisors of t. The following Theorem shows that $\tau(t)$ is of the form $\ln t+2\gamma-1$, or $O(\ln t)$.

Theorem 3.4 ([10]) $\sum_{i\leq t}\tau(i)=t\ln t+(2\gamma-1)t+O(\sqrt{t})$, where γ denotes Euler's constant.

(The Euler's constant is defined as $\lim_{n\to\infty} a_n$, where

$a_n=1+1/2+1/3+\ldots+1/n-\ln n$.)

Theorem 3.5 For any prime power q>2 and even integer k there exists an $(L,N,D)_q$-ECC, Γ, in which

1. $L=q^k$, $N=q(q^k-1)\tau(k/2)$;
2. D satisfies (1) with c given in (10).

4. EVALUATING THE CONSTRUCTION

Efficiency parameters of a sequential c-tracing schemes are (i) the size of the alphabet set, q, (ii) the length of the code, L, (iii) the number of users, N, and (iv) the number of colluders, c.

For sufficiently large q, a sequential tracing scheme obtained from a Reed-Solomon code has c approximately equal to $\sqrt{(q-1)/(k-1)}$. The number of codewords in RS-codes is given by q^k and so for fixed q, with an increase in k, the number of codewords increases while the value of c decreases. In other words since for a fixed c we have k approximately equal to $(q-1)/c^2+1$, if q is fixed the value of c determines k and hence to the number of codewords. To increase the number of codewords for a given value of c, q must be increased. However increasing q means that more information needs to be stored in each segment of the content. In watermarking algorithms the amount of embedded information is inversely proportional to the quality of the resulting marked content and as more information is embedded the quality of the marked content is reduced. So to maintain reasonable quality, larger segments must be used. This means that protection cannot be provided for shorter length content that cannot be broken into enough number of segments.

The code proposed in this paper overcomes this situation by allowing the number of codewords to be determined independent of c. For sufficiently large q, c is approximately equal to $\sqrt[4]{q}$ and so is independent of k. The number of codewords is $q(q^k-1)\tau(k/2)$ which increase with an increase in k and so for a fixed value of q, the number of codewords can be increased while keeping the level of security unchanged. This is a very important property and makes the code a flexible construction for a wide range of parameters.

ACKNOWLEDGEMENT

The authors would like to thanks Takeyuki Uehara for pointing out mistakes in the manuscript.

REFERENCES

[1] N.Alon and J.Spencer, *The probabilistic method*, Wiley, 1992.

[2] D.Boneh and M.Franklin, "An efficient public key traitor tracing scheme," *Advances in Cryptology - CRYPTO'99, LNCS* vol.1666, pp.338-353. Springer-Verlag, Berlin, Heidelberg, New York, 1999.

[3] B.Chor, A.Fiat and M.Naor, "Tracing traitors," *Advances in Cryptology - CRYPTO'94, LNCS* vol.839, pp.257-270. Springer-Verlag, Berlin, Heidelberg, New York, 1994.

[4] B.Chor, A.Fiat, M.Naor and B.Pinkas, "Tracing traitors," *IEEE Transactions on Information Theory*, Vol.46, No.3:893-910, 2000.

[5] A.J.Menezes (editor), *Applications of finite fields*, Kluwer Academic Publishers, 1993.

[6] A.Fiat and T.Tassa, "Dynamic traitor tracing," Advances in Cryptology - CRYPTO'99, LNCS vol.1666, pp.354-371. Springer-Verlag, Berlin, Heidelberg, New York, 1999.

[7] K.Kurosawa and Y.Desmedt, "Optimum traitor tracing and asymmetric schemes'" *Advances in Cryptology - EUROCRYPT'98, LNCS* vol.1462, pp. 502-517. Springer-Verlag, Berlin, Heidelberg, New York, 1998.

[8] R.Lidl and H.Niederreiter, *Finite fields*, Addison-Wesley Publishing Company, 1983.

[9] B.Pfitzmann, "Trials of traced traitors," *Information Hiding, LNCS* vol.1174, pp.49-64. Springer-Verlag, Berlin, Heidelberg, New York, 1996.

[10] H.E. Rose, *A Course in Number Theory*, Oxford, Clarendon Press, 1994.

[11] R.Safavi-Naini and Y.Wang, "Sequential traitor tracing," *Advances in Cryptology - CRYPTO 2000*, LNCS vol.1880, pp.316-332. Springer-Verlag, Berlin, Heidelberg, New York, 2000.

[12] R.Safavi-Naini and Y.Wang, "New results on frameproof codes and traceability schemes," *IEEE Transactions on Information Theory*, Vol. 47, No.7:3029-3033, 2001.

[13] R.Safavi-Naini and Y.Wang, "Sequential traitor tracing," Submitted, 2002.

[14] D.Stinson and R.Wei, "Combinatorial properties and constructions of traceability schemes and frameproof codes," *SIAM Journal on Discrete Mathematics*, 11:41-53, 1998.

[15] D.R.Stinson and R.Wei, "Key preassigned traceability schemes for broadcast encryption," *Proceedings of SAC'98, LNCS* vol.1556, pp. 144-156. Springer-Verlag, Berlin, Heidelberg, New York, 1999.

[16] M.A.Tsfasman and S.G.Vladut, *Algebraic-geometric codes* Kluwer Academic Publishers, 1991.

[17] Y.Wang, "Contributions to traceability schemes," *Ph.D Thesis*, School of Information Technology and Computer Science, University of Wollongong, Australia, 2001.

[18] J.Wolfmann, "The number of points on certain algebraic curves over finite fields," *Communications in Algebra*, 17(8):2055-2060, 1989.

[19] J.Wolfmann, "Algebraic curves and varieties over finite fields and irreducible cyclic codes," *Finite fields, coding theory, and advances in communications and computing, Lecture Notes in Pure and Applied Mathematics*, vol.141, pp.217-225. M. Dekker Inc. New York, 1993.

Chapter 17

SOFTWARE-DEFINED ANALYZER OF RADIO SIGNALS

Jerzy Lopatka

Communications Systems Institute, Military University of Technology, Warsaw, Poland

e-mail: jlopatka@wel.wat.waw.pl

Abstract: This chapter presents a scalable modular software receiver for technical analysis of radio emissions. It enables acquisition, recording, recognition and demodulation of various types of radio signals. Its construction is based on Texas Instruments TMS 320c67 digital signal processors.

Key words: software radio, signal analysis, modulation recognition

1. INTRODUCTION

Software radio is a concept of a dedicated computer system, that can be programmed to perform functions corresponding to the current user needs in a wireless communications area [1], [2]. Such a radio can be used as a general purpose equipment to achieve interoperability with a range of communications systems, local area networks, warning systems etc. Its construction makes it also suitable for modern radio monitoring systems, surveillance systems and electronic warfare purposes.

The proposed modular software receiver (MSR) is based on digital signal processing circuits. Its hardware consists of dedicated PCI or CPCI boards, which are used for analogue signal conditioning, converting into digital form, filtration and decimation. Main tasks, related to signal analysis, recognition and demodulation, are performed by DSPs. Frequency range, bandwidth of analysis, mode of operation and processing power of the radio can be achieved by appropriate hardware configuration. The CPCI version is dedicated for ruggedized solutions.

The platform is scalable and flexible, and can be used for various applications. The elaborated radio analysis software is also a modular one. Its functions and performance depend on the installed hardware and current needs.

2. HARDWARE PLATFORM

The MSR consists of five functional blocks [3](Fig. 1):

- analog signal conditioning block with ADC, working at 65 MHz sampling frequency;
- narrowband digital tuners with bandwidth up to 2 MHz;
- wideband digital tuners with bandwidth up to 25 MHz;
- a delay line with up to 512 MB RAM capacity, that gives delay up to 2 seconds;
- DSP boards equipped with up to 4 DSPs each.

All these blocks are implemented in three types of boards: analogue signal conditioning, ADC and narrowband tuner board (NTB), delay line and wideband tuner board (WTB) and DSP board. All these boards can work synchronously and they are interconnected by four types of buses:

1. *ADC bus* - 12/24 bits wide parallel bus, with clock frequency 65 MHz. The lower part of the bus is used for transmission of the output signal directly from ADC. The signal can be used e.g. by several tuners of the multi-channel receiver. The upper part of the bus can be used for transmission of the delayed ADC signal. It is necessary for searching receivers, monitoring solutions etc.
2. *Coreco's Auxiliary BUS (CAB)* - 32-bit bus working at 50 MHz. It is a basic bus for communications between tuners and DSP boards, and between different DSP boards.
3. *Serial buses* are the 50 MHz dedicated links between specified DSPs and other boards. They can be used for remote control, receiver tuning, AGC control etc.
4. *PCI or CPCI bus* is mainly used for boards programming purposes, but it can also be used either for received signals recording on HDD, or playback of previously recorded signals

MSR hardware can be configured according to the requirements. In the simplest case, if we want to process only simple, fixed frequency narrowband signals, the MSR contains only the NTB board. An input analogue signal is converted into digital form, mixed with local oscillator, decimated and filtered. The output signal is transmitted through the PCI interface to the PC memory and processed by the PC.

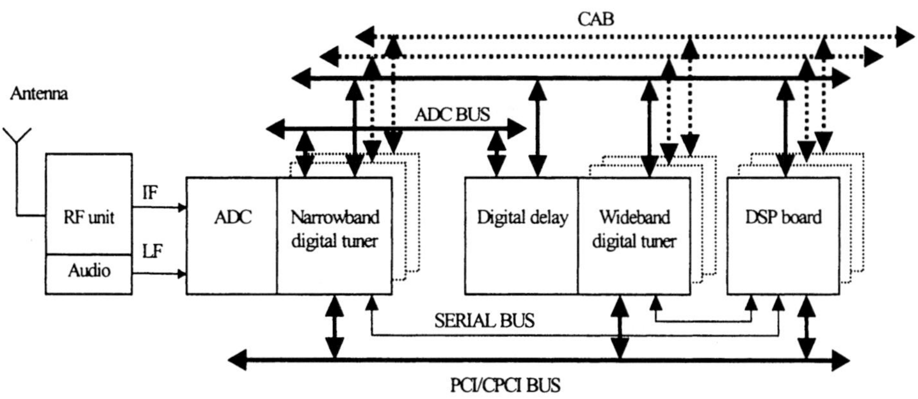

Figure 1. Block diagram of MSR

For more complex cases, the DSP board provides additional computational power. Downconverted signal is transmitted to the DSP using CAB. DSP can also control amplifiers, filters and the tuner, using dedicated serial bus.

The WTB board can be used for reception of the wideband signals. Because of the high data rate on its output, it works with one or more DSP boards.

MSR can also work as a combined search and reception receiver. In this case, the assumption is that we don't know signal parameters, modulation type, frequency range etc. For this purpose typical configuration consists of NTB, WTB and DSP boards. The wideband receiver is used for signal detection. DSP procedures evaluate signal bandwidth, centre frequency etc. These parameters are then passed to the NTB and the narrowband tuner is tuned to the desired signal. Then DSP can analyse modulation type, signal parameters, data rate etc. and perform its demodulation. In this configuration NTB uses a delayed signal from the ADC bus, to compensate computational time of detection procedures. This configuration of MSR enables a tracing detection of frequency hopping signals, CHIRP signals and BURSTs reception.

For multi-channel applications it is also possible to use multiple boards of given type. They work synchronously to avoid signal inter-modulations.

3. NARROWBAND TUNER BOARD

The MSR can process either analogue or digital signals (Fig. 2). It can use two signals from the external receiver: high or intermediate frequency signals (IF, RF) and audio signals (LF). The first one is used for analysis and demodulation of analogue and digitally modulated signals. It is also suitable for selection of particular channels from the FM signals with frequency divided multiplex (FM-FDM), and analysis of signals with secondary FM modulation. The IF signal can be either analogue or digital IQ. The audio input can be used for analysis of low frequency demodulated signals, wired modems etc.

Analogue signals are conditioned using low-pass filters with 30 MHz and 50 kHz bandwidth respectively, and two adjustable amplifiers. The amplifiers are controlled by DAC and can be used for analogue gain control. Analogue to digital conversion is performed with accuracy of 12 bits and sampling frequency of 65 MSps. Digital signal from the ADC is transmitted to the ADC bus and the narrowband digital tuner (HSP 50214). The tuner contains tuneable quadrature digital local generator, complex mixer and programmable decimating filters [4]. The filters are FIR filters that provide ideally linear phase characteristics. Output signal can be phase-adjusted using poli-phase interpolator. Signal on the output of the down-converter is a complex signal with zero centre. An overall decimation factor of the tuner D is given by:

$$D = D_{CIC} \cdot D_{HB} \cdot D_{FIR} \qquad (1)$$

where: D_{CIC} - decimation factor of CIC filters – $2^2 \ldots 2^5$,
D_{HB} - decimation factor of half-band filters – $2^0 \ldots 2^5$,
D_{FIR} - decimation factor of FIR filters – $2^0 \ldots 2^4$

It corresponds to the signal bandwidth:

$$B = f_s / D \cdot F \qquad (2)$$

where: f_s - sampling frequency,
F - FIR bandwidth normalized in range of [0...1].

Maximum decimation factor is equal 16384, which gives minimal sampling frequency about 4 kHz. Maximum bandwidth cannot exceed about 2 MHz.

The tuner also contains frequency discriminator that can work in phase locked loop, or as a general purpose FM demodulator.

Optional second tuner is used for the FM signal analysis. In this case the first tuner works as a FM demodulator and its output signal is a source of data for the second tuner that extracts particular channel from the FM-FDM signals. Output signal can be transmitted by CAB to DSP for analysis and through the PCI to the PC for storage. Stored signals can also be played back from the HDD and transmitted to the DSP for multiple analysis. Serial buses are used by the DSP for remote control of tuners and the DAC.

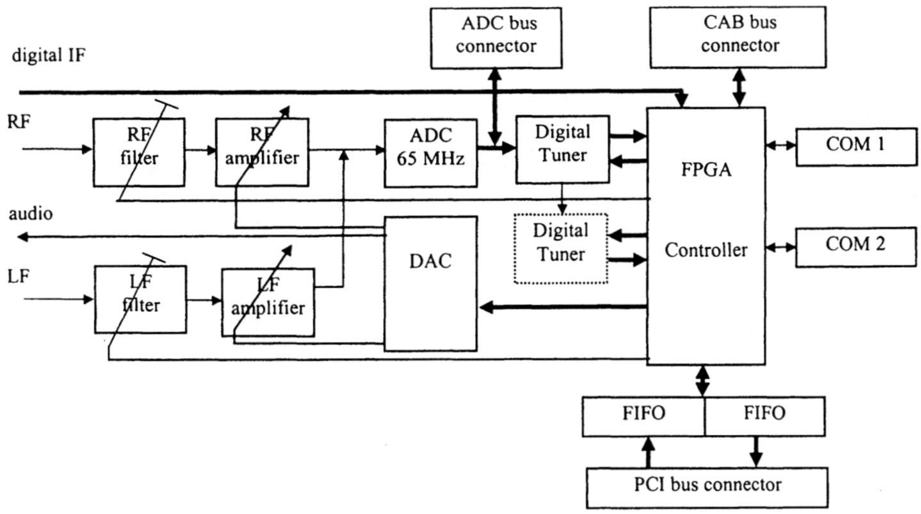

Figure 2. Block diagram of the narrowband tuner board

4. WIDEBAND TUNER BOARD

The WTB board consists of two functional blocks: the delay line and the tuner. Input signal is a 12-bit digital signal from the ADC bus (Fig. 3) and it can be delayed using a "swing buffer". The buffer consists of at least 2 RAM blocks. These blocks are alternatively used for writing and reading operations. The shift between write and read pointers is always the same and is equal to the block size. The input data clock is fixed, so the block size determines also the maximum delay. To achieve a variable delay, the FPGA controller uses only a selected area from each memory block to store the data. The delayed signal can be processed by the wideband tuner or transmitted back to other boards (e.g. narrowband tuner) using upper half of the ADC bus. Maximum size of memory is 256Mx16 and it gives a delay up to 2 seconds.

The wideband tuner (GC1012A) is a digital down-converter [5]. Signals are converted to the baseband, filtered using low-pass filters and decimated. The decimation factor value is programmed at rates 2...64. The decimator can also be skipped. The pass-band of the output filter covers about 80% of the output bandwidth. The output signal can be a complex or real one, whereas the spectrum can be also inverted or shifted by a half of the output sampling rate.

The narrowband tuner (HSP 50214) is an optional circuit. It can work as the second tuner, after the signal is pre-processed by the wideband tuner or they can work in parallel way. If the wideband tuner works as a front–end mixer and filter, its output signal can be configured as a real one. The signal spectrum is then shifted by a half of the output sampling rate and signal can be fed to the input of the narrowband tuner, that performs further filtration and decimation. This configuration is useful both for analysis of very narrowband signals, when the decimation factor of a single tuner is insufficient and for reception of signals in presence of strong adjacent interfering signals where high selectivity is required.

In the parallel configuration, the wideband tuner is used for wideband signal reception, whereas the other one is used for frequency offset tracking, CHIRP signals carrier frequency detection or tracking reception of FH signals. In this case the narrowband tuner can use a delayed input signal to compensate the calculation time in the DSP.

Output signals can be transmitted through CAB to the DSP for further analysis. The PCI interface is used mainly for board configuration. The serial bus is used for remote control of tuners and delay line by the DSP.

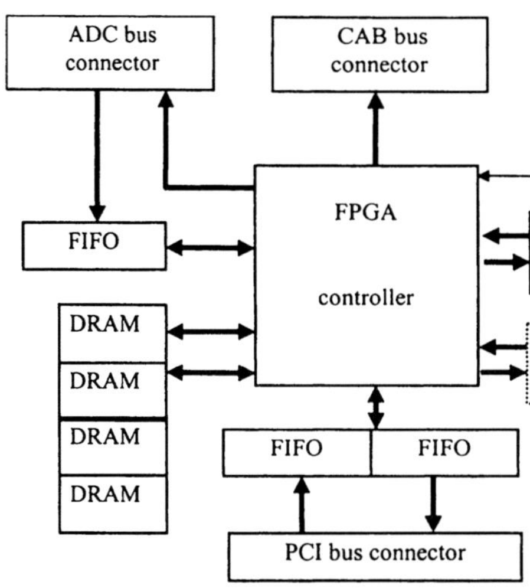

Figure 3. Block diagram of the wide-band tuner board

5. DIGITAL SIGNAL PROCESSORS BOARD

The Python/C67 PCI/CPCI board is a multi-DSP processing card exploiting Texas Instrument's TMS320C6701 floating point DSP (Fig. 4) [6]. The Python/C67 supports up to four TMS320C6701 DSPs interconnected via high-speed communication links allowing data to be communicated quickly between processors. This high-speed architecture is effective for implementing either pipelined or parallel image and signal processing applications. The Python/C67 is an expansion capable and connected to other boards through the 32-bit CAB (Coreco Auxiliary Bus) at 50 MHz. Key features of the board are:

- up to four TMS320C6701s working at 167 MHz;
- 4 GFLOPs (peak) processing power on a single board;
- 512 KB of SBSRM per DSP;
- 16 MB of SDRAM per DSP;
- 4 MB of shared EDORAM;
- direct inter-DSP communication links;
- on-board interface to PMC module;
- 32-bit CAB bus interface at 50 MHz;
- integrated JTAG controller;
- on-board Intel i960 controller.

The board is very flexible and provides fast communications interfaces that make it suitable for real time applications. CPCI version is software compatible with PCI version that enables easy upgrade to the ruggedized version of MSR.

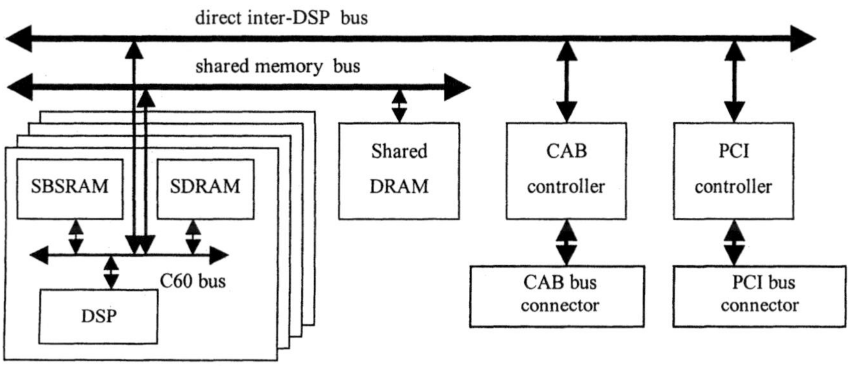

Figure 4. *Block diagram of CORECO PYTHON C67 board*

6. SOFTWARE STRUCTURE

MSR software is a modular multilevel application. The lowest level realizes boards configuration, data transfers, remote control, tasks management etc. and is invisible for the user. Higher-level modules perform signal acquisition, analysis, detection, recognition, demodulation and decoding. The highest-level software provides interactive graphic user interface.

In the designed software, signal analysis can be performed either manually or in automated way. To determine signal bandwidth and carrier frequency FFT, mean FFT, maximum FFT and waterfall calculations are performed. Real and imagine parts of the signal, its constellation, momentary magnitude, phase, and frequency time plots are also available. Precise parameters measurement is also possible on the basis of histograms and eye patterns. Structure of the transmitted data can be analysed exploiting facsimile display, where each data bit is depicted as a horizontal line. For plain transmissions it enables determination the length of bytes, packets structures and protocols. It makes possible an extraction of useful data. Received bit flow can be decoded using user defined decoder, that can reject synchronisation words, split bits into several channels and decode them according to specified transmission alphabet [7].

6.1 Signal Recognition

Signal recognition is performed using DSP board. DSPs calculate signal statistics, and then use them as distinctive features for modulation scheme classification.

During the initial step, a parametric, autoregressive (AR) model of the signal is used for spectral identification because AR spectral estimates have sharp peaks that correspond to carriers in radio signals, and Gaussian-like components reflecting the modulated bands. These methods are computationally effective and allow creating adaptive AR filters as models of analysed signals. To achieve precise carrier frequency estimation and to avoid spectral peaks splitting, the LS method was chosen [8].

In real conditions, the signal carrier frequency can be shifted because of transmitter frequency offset, inaccurate receiver tuning or Doppler shift. Such a signal should be also recognized properly, and the frequency shift should be also determined. It means that relations between parameters of original and its shifted version should be easy to determine, the shift should be easy to calculate and some simple transformations should correct the shift.

These requirements are met by the poles co-ordinates of the AR filter [9]. In this case filter transfer function is given by:

$$H(z) = \frac{A}{\prod_{k=1}^{p} (1 - d_k z^{-1})} \tag{3}$$

where: d_k denote the poles locations.

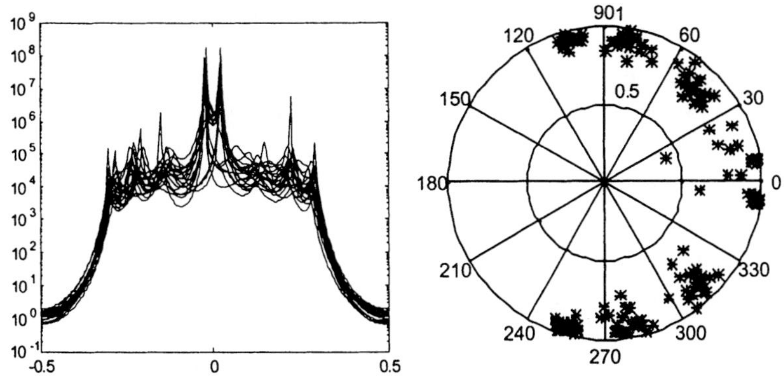

Figure 5. FSK signal spectrum estimate and poles locations for AR model order 8

Poles' internal dissipation is low enough, especially for poles reflecting signal components with high energy (Fig. 5), and in case of frequency shift, the poles are simply rotated around the begin of co-ordinates system. The angle shift of all poles is the same and poles' magnitudes remain unchanged.

On the basis of poles' location on the z plane, after some additional processing, pattern recognition is performed using neural network. It is a 2 layer, 60 neurons network with log-sigmoid transfer function. Results of this method allows mainly on FSK-MFSK-PSK distinction. It also enables to determine the centre frequency shift. Further analysis of radio signal parameters is performed on the basis of signal statistics, which are used as distinctive features. A fuzzy logic test is used for modulation scheme classification [10]. According to selected signal type, basic parameters as data rate, carrier frequency and frequency shift are also calculated. All these parameters are used for demodulator selection and its proper configuration.

6.2 Signal Demodulation

To provide signal processing in a real time, the demodulation procedures are realized using DSPs. Single DSP board enables real time demodulation of the 25kHz channel. Demodulator parameters are assigned by PC, through the PCI interface. The demodulation results are also transmitted to the PC. The IQ, AUDIO and LINE signals are also available in an analogue form for scope analysis, listening to the signal and connecting to another external devices.

In the DSP board the following demodulators are implemented:

- discriminator based FSK,
- correlation based FSK,
- M-ary discriminator based FSK,
- 2 PSK,
- 2,4,8 DPSK version A,

- 2,4 DPSK version B,
- 16 QAM.

The realized algorithm depends on the selected type of demodulator and the appropriate procedures are performed in a sequence (Fig. 6).

6.2.1 Lowpass filter

It is a fourth order Chebyschev type II filter. Real and imagine parts of the signal are filtered separately, so the filter acts as a passband filter, with cut-off frequencies equal 0.1,0.2,..0.9 of the downconverter bandwidth. It allows suppression of neighbour-channel interferences and better filtration of the useful signal.

6.2.2 Frequency discriminator

This procedure performs calculation of instantaneous frequency using complex *Re(n)* and *Im(n)* components of the signal *x(n)*. It is used for discriminator based FSK and MFSK demodulators. The frequency is estimated on the basis of the following formula:

$$f(n) = \frac{[\phi(n) - \phi(n-1)] \bmod \pi}{2\pi} f_s \tag{4}$$

where: $\phi(n) = arctg(\mathrm{Im}(n) / \mathrm{Re}(n))$;

To decrease the time of computations, a table of *arctg* function is calculated once and placed in RAM.

6.2.3 Fitted filters

Fitted filters are used in correlation based FSK demodulator. They are performed by complex mixing of the signal *x(n)* with *k* selected nominal frequencies. The output signal is achieved using sum and dump operation:

$$u_{Fk}(n) = \sum_{n=0)}^{N-1} x(n-l)\sin(2\pi f_k l / f_s) \tag{5}$$

where: f_k - *k-th* nominal frequency;
N - symbol length.

Real time constrains caused that *sinus* table is calculated once and placed in DSP RAM. Reading of succeeding samples is performed by a set of pointers incremented with selected step and offset.

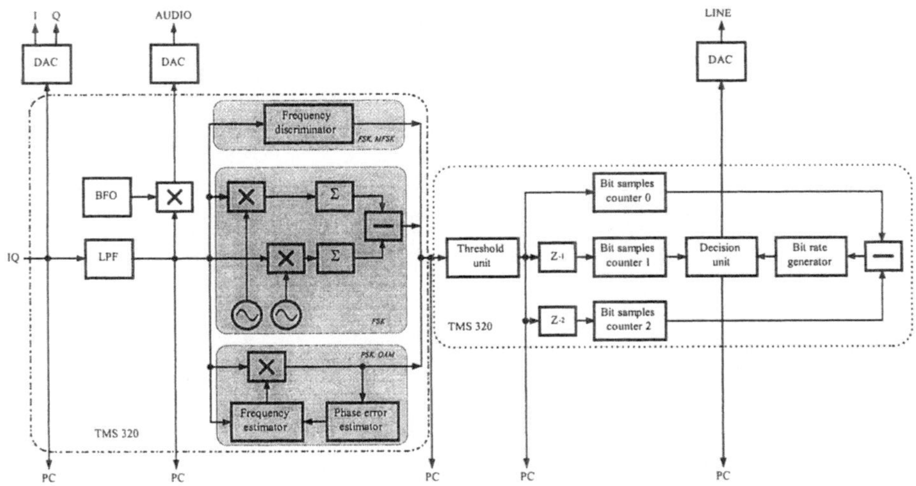

Figure 6. Block diagram of the demodulator software

6.2.4 Carrier frequency and phase recovery

Carrier recovery is performed in PSK, DPSK and QAM demodulator. These procedures are almost identical. The only difference lies in the threshold level and time constant.

Average frequency is calculated according to the equation:

$$f_A(n) = \begin{cases} \alpha \cdot f_A(n-1) + (1-\alpha) f(n): & \text{mod}(n) > \text{mod}_0 \ \textit{and} \ |f(n) - f_A(n-1)| < f_0 \\ f_A(n-1) & : \quad \textit{other cases} \end{cases} \tag{6}$$

where: α - time constant,

$\text{mod}(n) = \sqrt{\text{Re}(n)^2 + \text{Im}(n)^2}$,

mod_0 - threshold value of the amplitude,

f_0 - threshold value of the frequency.

Carrier phase recovery is performed in a loop, where the phase error is an average difference between received phase and expected points of constellation. Average phase error is calculated as follows:

$$\varphi_A(n) = \begin{cases} \beta\varphi_A(n-1) + (1-\beta)(\varphi(n) - \varphi_C(n)): & \text{mod}(n) > \text{mod}_0 \ \textit{and} \ |\varphi(n) - \varphi_C(n)| < \varphi_0 \\ \varphi_A(n-1) & : \quad \textit{other cases} \end{cases} \tag{7}$$

where: β - time constant,

φ_C - phase of the closest point of constellation,

mod_0 - threshold value of the amplitude,
φ_0 - threshold value of the phase error.

6.2.5 Threshold unit

This unit works as a Schmitt relay. It has selected *on* and *off* thresholds, according to centre frequency, frequency shift etc. In case of QAM demodulator, this unit provides also accurate AGC control and determination of constellation position.

6.2.6 Time base recovery

Time base recovery is based on three counters method. Accurate counter output is used for decision unit, whereas outputs from delayed and advanced counters are compared and the error is used to correct the time base.

6.2.7 Decision unit

This unit acts as a comparator in case of 2-value modulations and as a quantizer for *m*-ary modulations. Thresholds values and their number can be determined by PC.

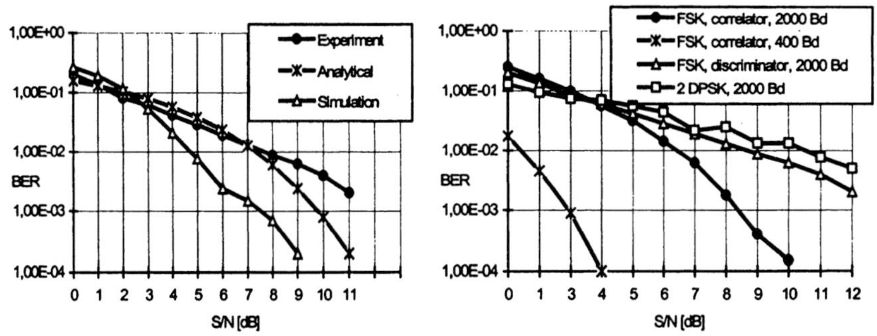

Figure 7. System performance
a) Comparison of simulation and experimental results for FSK,
b) BER of selected demodulators versus signal to noise ratio.

6.2.8 System performance

Using developed test stand, some performance tests were performed. The first step was a validation of achieved results. It was done for all demodulators. It is shown in Fig. 7a a FSK discriminator demodulator performance comparison for: analytical BER, experimental results and results achieved from MATLAB based demodulator model, identical to DSP algorithm. The difference between the latter two curves shows an impact of finite precision calculations in DSP and influence of real radio receiver, and downconverter.

It is shown in Fig. 7b that correlation based FSK demodulators are less sensitive to jamming than the frequency discriminator based one. This advantage is larger for slower transmissions. Fitted filters of the demodulator have narrow bandwidth that eliminates most of the jamming signal. Discrimination based demodulator has no implemented filters, so performances are similar to PSK demodulator. The achieved results correspond to theoretical values that confirm proper work of the MSR.

7. CONCLUSIONS

The designed modular receiver is a flexible platform for multilevel analysis and reception of radio signals. Its performance and computational power can be adjusted according to current user needs. Modular software enables implementation of both wideband and narrowband signals analysis. It can be also used in advanced radio-monitoring systems, EW applications etc. The MSR can be also supplemented with digital up-conversion board to create a fully operating software radio.

REFERENCES

[1] J. Place, D. Ker and Rd. Schafer, "Joint Tactical Radio System," *Milcom 2000*, pp. 209-214.

[2] J. Mitola III, "SDR Architecture Refinement for JTRS," *Milcom 2000*, pp. 214-219.

[3] Lopatka Jerzy, "Modular Software Receiver for Radio Signal Analysis," *6th International Symposium on DSP for Communication*, Sydney-Manly 28-31 January 2002, pp. 219-224.

[4] Intersil Corp., "HSP 50016 datasheet," June 1999.

[5] Graychip Inc., "GC1012A –digital tuner chip datasheet," February 1998.

[6] Coreco Inc., "PYTHON/C6 – datasheet," 1998.

[7] J. Lopatka, "Adaptive Multiwaveform Demodulator Of Radio Signals," *5th International Symposium on Digital Signal Processing for Communication*, Perth, 1999, pp. 37-43.

[8] L. Marple, *Modern Spectrum Estimation*, Prentice-Hall International Inc., 1988.

[9] J. Lopatka, "Recognition of Narrowband Radio Signals Using Autoregressive Models and Pattern Comparison Approach," *Journal of Telecommunications and Information Technology*, No 1/2002, pp. 56-59.

[10] J. Lopatka and M. Pedzisz, "Fuzzy logic classifier for radio signals recognition," *Journal of Telecommunications and Information Technology*, No. 2/2001, pp. 31-35.

Chapter 18

INTERLEAVED PC-OFDM TO REDUCE PEAK-TO-AVERAGE POWER RATIO

A. D. S. Jayalath and C. Tellambura
School of Computer Science and Software Engineering, Monash Univerisy, Australia

Abstract: Parallel combinatory orthogonal frequency division multiplexing (PC-OFDM) yields lower maximum peak-to-average power ratio (PAR), higher bandwidth efficiency and lower bit error rate (BER) on Gaussian channels compared to OFDM systems. However, PC-OFDM does not improve the statistics of PAR significantly. In this chapter, the use of a set of fixed permutations to improve the statistics of the PAR of a PC-OFDM signal is presented. For this technique, interleavers are used to produce $K-1$ permuted sequences from the same information sequence. The sequence with the lowest PAR, among K sequences, is chosen for the transmission. The PAR of a PC-OFDM signal can be further reduced by 3-4 dB by this technique. Mathematical expressions for the complementary cumulative density function (CCDF) of PAR of PC-OFDM signal and interleaved PC-OFDM signal are also presented.

Key words: OFDM, Peak-to-average power ratio, PC-OFDM, Data permutation.

1. INTRODUCTION

Multicarrier modulation such as OFDM has been investigated and demonstrated as a robust technique for high speed data transmission severe multipath fading channels [1]. Moreover, OFDM transceiver can be implemented efficiently using the fast Fourier transform algorithm. OFDM applications include digital audio broadcasting, asynchronous digital subscriber lines, digital video broadcasting and wireless local area networks. Due to the nonlinearities of transmitter power amplifiers, high PAR values of the OFDM signal generate out-of-band noise (OBN). Further more nonlinearities of amplifier cause inband distortions of the signal giving higher BERs. Back-off required at the amplifier can be reduced by designing OFDM systems with low PAR.

Solutions to the PAR problem of OFDM have been proposed by several researchers [2, 3]. PC-OFDM [4] reduces the absolute maximum PAR. PC-OFDM is generated by inserting zero amplitude subcarriers into an OFDM signal, reducing the number of active subcarriers. To increase the system throughput a set of bits called parallel combinatory bits are mapped to zero amplitude positions. PC-OFDM can carry more information bits than a similar OFDM system depending on the number of zero amplitude subcarriers. The reduction in PAR is measured in terms of the theoretically maximum, $10\log_{10}(N)$, where N is the number of active carriers. Despite increasing the number of zero subcarriers, PC-OFDM offers negligible improvements in the statistics of the PAR. However, statistical properties of PC-OFDM can be improved further by using other PAR reduction methods.

In this chapter, the use of data permutation (interleaving) to improve the statistics of PAR of PC-OFDM signals is investigated. Interleaved PC-OFDM is obtained by combining the data permutation technique reported in [5] with PC-OFDM. Large PAR reductions can be achieved by interleaved OFDM at the expense of system complexity. The PAR statistics of an OFDM signal is evaluated by using the complementary cumulative distribution function (CCDF = $\Pr(\text{PAR}>\text{PAR}_0)$. Mathematical expressions to evaluate the CCDF of PC-OFDM signal and interleaved PC-OFDM signal are also presented.

In Section 2 the OFDM system model and the theoretical description of the system are given. Section 3 presents the non-lineartransmitter power amplifier characteristics. The simulation and theoretical results are presented in Section 4. Section 5 concludes the chapter.

2. SYSTEM MODEL

2.1 An OFDM system and PAR

A block of N symbols, $\mathbf{X}=\{X_1,X_2,\ldots,X_{N-1}\}$, is formed with each symbol modulating one of a set of N subcarriers with frequency, $f_n, n-0,1,\ldots,N-1$. The N subcarriers are chosen to be orthogonal, that is $f_n = n\Delta f = n/T$, where and T is the OFDM symbol duration. The complex baseband signal can be expressed as

$$x(t)=\frac{1}{\sqrt{N}}\sum_{n=0}^{N-1}X_n e^{j2\pi f_n t},\quad 0\le t\le T \tag{1}$$

This signal can be generated by taking an N point inverse discrete Fourier transform (IDFT) of the block $\mathbf{X}$ followed by low pass filtering. The actual transmitted signal is modeled as real$\left[x(t)e^{j2\pi f_c t}\right]$, where f_c is the carrier frequency. The PAR of the transmitted signal in (1) can be defined as

$$\xi=\frac{\max|x(t)|^2}{E[|x(t)|^2]} \tag{2}$$

where $E[x]$ is the expected value of x. The PAR of the continuous-time OFDM signal cannot be computed precisely by the use of the Nyquist sampling rate [6], which amounts to N samples per symbol. In this case, signal peaks are missed and P AR reduction estimates are unduly optimistic. Oversampling by a factor of 4 is sufficiently accurate and this is achieved by the simple computation of the $4N$-point IDFT of the data frame. The maximum of the PAR for an N subcarriers OFDM system is N. Therefore we can reduce the maximum by setting some subcarriers to zero. This is what motivates PC-OFDM.

2.2 PC-OFDM

Parallel combinatory OFDM signals reported in [4] can be described as follows. For $N=N_{\text{null}}+N_{\text{pc}}$, consider N_{pc} subcarriers with M-PSK signal points and N_{null} subcarriers with zero amplitude points. This gives a system with the maximum PAR of $10\log_{10}(N_{\text{pc}})$ dB, which is lower than the maximum PAR of an OFDM system with N subcarriers. This new system is called a parallel combinatory OFDM system. In the transmitter we first choose which subcarriers to be zero and nonzero respectively. Then N_{pc} subcarriers are modulated by M-phase shift keying (M-PSK) constellation points. Thus m_{psk}, $N_{\text{pc}}\log_2(M)$ bits are mapped to a subset of subcarriers used. Choosing N_{pc} out of N subcarriers can be done in $\binom{N}{N_{\text{pc}}}$ different ways. The total number of bits m_{tot} that can be transmitted using a PC-OFDM signal can be found as

$$m_{\text{tot}}=m_{\text{psk}}+m_{\text{pc}}=N_{\text{pc}}\log_2 M+\left\lfloor \log_2\binom{N}{N_{\text{pc}}}\right\rfloor \tag{3}$$

where $\lfloor x \rfloor$ is the largest integer smaller than, or equal to x.

Reference [4] suggests to use a maximum likelihood (ML) detector at the receiver after the FFT operation to recover the data. The ML-detector maps the signal points in the subcarriers to the closest signal point in the ($M+1$)-APSK (amplitude and phase shift keying) constellation. If exactly N_{pc} subcarriers are mapped to signal points with non-zero amplitudes, the received symbol is accepted and decoded. Otherwise, if there are more subcarriers with zero amplitude, $N_{null}+L$, L subcarriers close to any of the nonzero constellation points are decoded to the corresponding non zero point in the ($M+1$)-APSK constellation. If the number of zero subcarriers is less than N_{null}, $N_{null}-L$, L nonzero subcarriers having the smallest amplitudes are set to zero. Received bits are decoded after these corrections.

2.3 $M+1$-APSK Signal constellation

The ($M+1$)-APSK signal constellation is constructed from an M-PSK constellation. There is a signal point at the origin of the signal space. The amplitudes of the other M constellation points depend on the N/N_{pc} ratio, $\gamma=\sqrt{N/N_{pc}}$. Therefore, for each PC-OFDM system we have to define a unique signal constellation. Figure 1 shows the 9-APSK signal constellation.

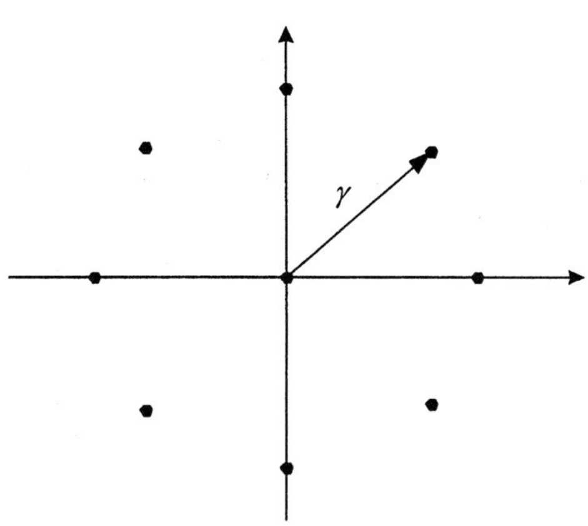

Figure 1. Signal constellation for 9-APSK

The bandwidth efficiency of PC-OFDM may be higher than OFDM. The problem of mapping bits to PC-OFDM signal is solved by using the Jonson association scheme, together with a position algorithm for the PSK symbols [4].

2.4 Interleaver approach

The interleaver approach to reduce PAR of an OFDM signal is presented in [5]. In this approach $K-1$ interleavers are used at the transmitter. These interleavers produce $K-1$ permuted frames of the input data before mapping symbols. The four times oversampled IDFT of each frame (including the original frame) is used to compute its PAR. The minimum PAR frame of all the K frames is selected for transmission. The identity of the corresponding interleaver is also sent to the receiver as side information. A similar approach has been independently developed by Ochiai and Imai [7].

2.5 Theoretical CCDF of PAR of PC-OFDM

Reference [8] provides an empirical formula for the distribution of the PAR. Reference [7] derives an expression for the exact distribution of the PAR. As this expression is complex and difficult to solve an approximation of the peak distribution based on level-crossing rate is also presented. This simple expression is used here to predict the PAR distribution of interleaved PC-OFDM signals.

The CCDF of PAR of an OFDM signal can be expressed as

$$\Pr(\xi > \xi_0) \approx \Pr(\xi > \overline{\xi} \mid \xi_0 > \overline{\xi}) = \begin{cases} 1-\left(1-\dfrac{\xi_0 e^{-\xi_0^2}}{\overline{\xi} e^{\overline{\xi}^2}}\right)^{\sqrt{\frac{\pi}{3}}N\overline{\xi}e^{-\overline{\xi}^2}} & \text{for } \xi_0 > \overline{\xi}, \\ 1 & \text{for } \xi_0 \le \overline{\xi}. \end{cases} \quad (4)$$

This was derived considering only the peaks exceeding a given threshold $\overline{\xi}$ above zero. A proper $\overline{\xi}$ is selected by making the assumption: each positive crossing of the level has a single positive peak that is above the level $\overline{\xi}$ [7]. It also suggests $\overline{\xi} = \sqrt{\pi}$ for QPSK modulation and slightly lower value for 16-QAM. Next, we derive an expression for the PAR distribution of PC-OFDM signals.

The PC-OFDM signal has fixed number of zero amplitude subcarriers. Using (4) we can express the CCDF of PAR of the PC-OFDM signal as follows:

$$\Pr(\xi > \xi_0) \approx \Pr(\xi > \bar{\xi} \mid \xi_0 > \bar{\xi}) = \begin{cases} 1 - \left(1 - \dfrac{\xi_0 e^{-\xi_0^2}}{\bar{\xi} e^{\bar{\xi}^2}}\right)^{\sqrt{\frac{\pi}{3}} N_{pc} \bar{\xi} e^{-\bar{\xi}^2}} & \text{for } \xi_0 > \bar{\xi}, \\ 1 & \text{for } \xi_0 \leq \bar{\xi}. \end{cases} \quad (5)$$

Now, for the interleaved PC-OFDM signal we assume that all K sequences are independent and uncorrelated. The validity of this assumption is confirmed via simulations. Then the CCDF of PAR of the interleaved PC-OFDM signal can be expressed as

$$\Pr(\xi > \xi_0) \approx \Pr(\xi > \bar{\xi} \mid \xi_0 > \bar{\xi}) = \begin{cases} \left[1 - \left(1 - \dfrac{\xi_0 e^{-\xi_0^2}}{\bar{\xi} e^{\bar{\xi}^2}}\right)^{\sqrt{\frac{\pi}{3}} N_{pc} \bar{\xi} e^{-\bar{\xi}^2}}\right]^K & \text{for } \xi_0 > \bar{\xi}, \\ 1 & \text{for } \xi_0 \leq \bar{\xi}. \end{cases} \quad (6)$$

3. NONLINEAR TRANSMITTER CHARACTERISTICS

The nonlinear distortion at the transmitter causes interferences both inside and outside the signal bandwidth. The inside component determines the amount of bit error rate degradation of the system, whereas the outside component effects the adjacent frequency band (OBN). The transmitter nonlinear distortion include signal clipping in the analog to digital (A/D) converter, signal clipping in the IFFT and FFT processors with a limited word length, amplitude modulation (AM)/AM and AM/phase modulation (PM) distortion in the radio frequency (RF) amplifiers. The OBN of OFDM signals increases due to nonlinear power amplifiers operating at lower back-offs. The high PAR of OFDM requires high back-offs at the amplifiers. The non-linear characteristic of the soft limiter is shown below [9]. The performance of PC-OFDM and a soft limiter is simulated in this chapter.

3.1 Soft limiter (SL)

Since the AM/PM component is zero the nonlinear characteristics of a SL can be written as

$$g(x) = \begin{cases} x & |x| \leq A \\ Ae^{j\phi} & |x| > A \end{cases} \tag{7}$$

where x is the input to the SL, A is the saturated output and ϕ is the phase angle of the input x. Although most physical components will not exhibit this piecewise linear behavior, the SL can be a good model if the nonlinear element is linearized by a suitable predistortor. The back off (BO) at the non-linear device can be defined in terms of maximum power output A^2 as

$$\text{BO}=10\log_{10}\left(\frac{A^2}{E[|x|^2]}\right) \tag{8}$$

For PSD results, it is convenient to define the normalized bandwidth $B_n = fT/N$ where T is the OFDM symbol duration.

4. RESULTS

PC-OFDM systems with $N = 32$, $N_{\text{pc}} = 12, 16, 20, 28$ and 5-APSK modulation are simulated. Figure 2 shows a comparison of theoretical and simulated CCDFs of PC-OFDM signals. The theoretical expression (5) approximates the simulation results well. Very little improvement of PAR statistics is observed when the number of zero subcarriers (N_{null}) in a PC-OFDM system is increased. Figure 3 depicts the CCDF of PAR of the PC-OFDM and interleaved PC-OFDM signals with $N_{\text{pc}} = 28$ and 12. No PAR reduction can be observed in PC-OFDM, when $N_{\text{pc}} = 28$ and there is about 0.5dB reduction in PAR at 0.1%PAR when $N_{\text{pc}} = 12$ (throughout the chapter we denote by *n*%PAR, a threshold, that PAR exceeds this threshold less than *n*% of OFDM blocks) compared to the PAR statistics of an OFDM system with $N_{\text{pc}} = 32$ subcarriers.

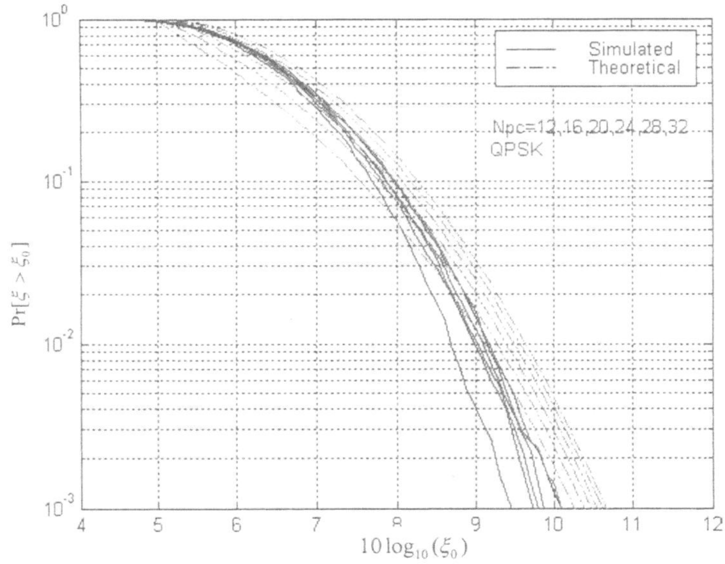

Figure 2. Theoretical and Simulated CCDF of PC-OFDM.

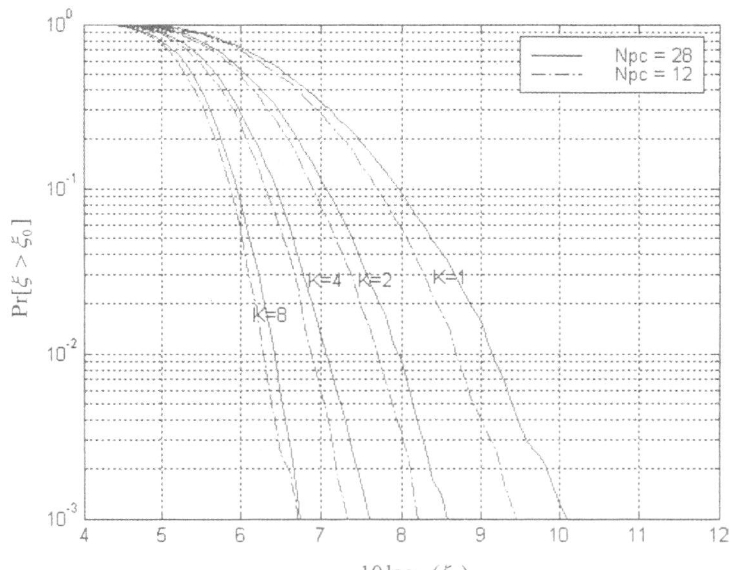

Figure 3. CCDF of PC-OFDM with interleaving to reduce PAR ($M = 4$)

Statistics of PAR is improved very slightly when the number of subcarriers in an OFDM system is reduced, although it gives a low value for the maximum possible PAR. For example, when $N = 32$ the maximum possible PAR is $10\log_{10}(32) = 15.1\,\text{dB}$, while this is 14.4dB when $N = 28$ although there is a 1dB reduction in the maximum value of PAR, the statistics remain unchanged above the region 0.1%PAR. For $N = 32$ and 12, the difference in maximum PAR is about 4.3dB, while the statistics show only 0.5dB reduction at 0.1%PAR. Use of interleaving will not improve the maximum PAR but will improve the PAR statistics. Total PAR reduction of 4dB is observed for N_{pc} when interleaving (number of interleavers are 15, $K = 16$) is used. A 1.5dB improvement of PAR statistics at 0.1%PAR is observed even with a single interleaver. Similar performance is observed for PC-OFDM with $N_{pc} = 12$.

Figure 4 and Figure 5 compare theoretical and simulated CCDFs of interleaved PC-OFDM for $N_{pc} = 28$ and 20. Theoretical expression (6) can therefore predict the PAR statistics of any interleaved PC-OFDM system. The complexity of a PC-OFDM system increases with the number of subcarriers. Therefore, the expressions (5) and (6) can predict the performance without performing extensive computer simulations.

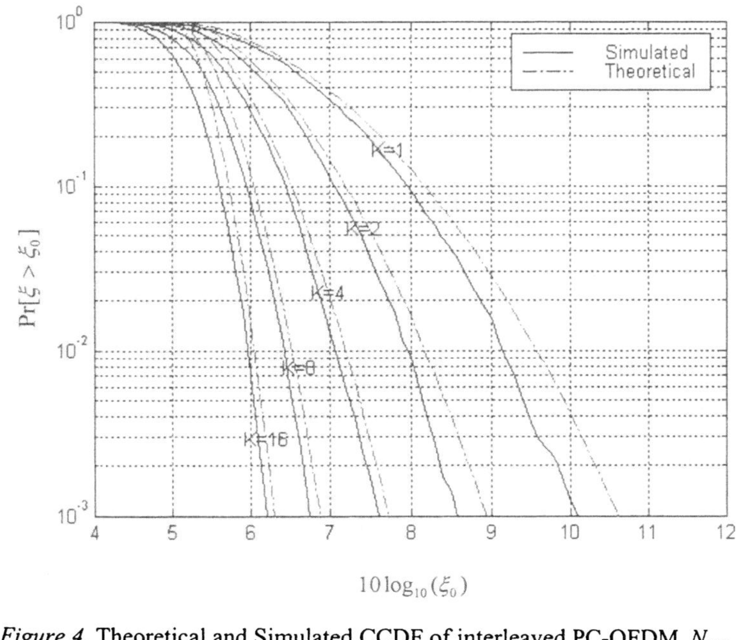

Figure 4. Theoretical and Simulated CCDF of interleaved PC-OFDM $N_{pc} = 28$

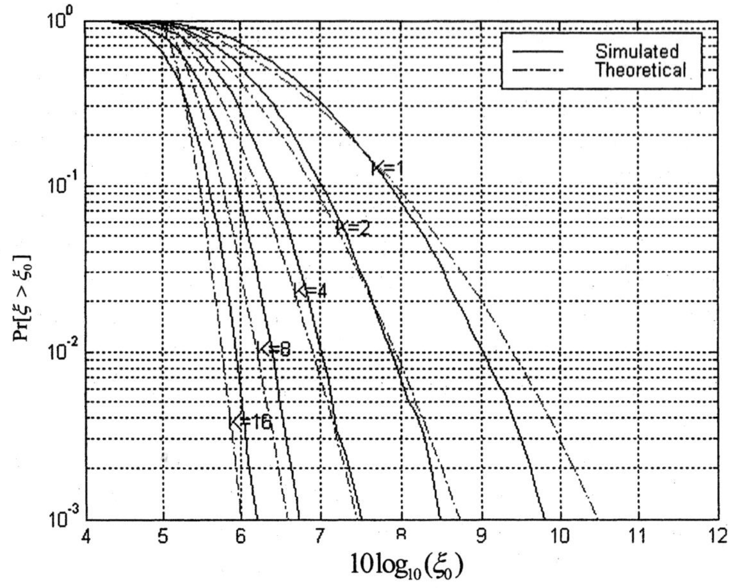

Figure 5. Theoretical and Simulated CCDF of interleaved PC-OFDM $N_{\text{pc}} = 20$

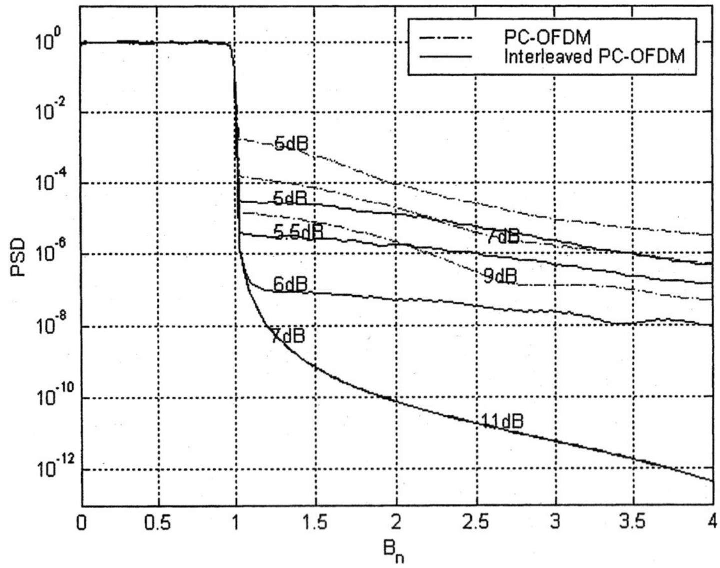

Figure 6. Power spectral density of PC-OFDM and interleaved PC-OFDM.

Figure 6 shows the power spectral density of the PC-OFDM signal when passing through a SL non-linearity. A clear reduction in OBN is observed in interleaved PC-OFDM systems. Simulation results are shown for an PC-OFDM system with $N_{pc} = 28$ and 5-APSK modulation. When the clipping level of SL is at 5dB a 10dB reduction in OBN is observed at $2B_n$, while the reduction is more than 50dB when the clipping level is 7dB. To achieve similar performances with PC-OFDM SL clipping level has to be increased to 11dB. Therefore, interleaved PC-OFDM reduces the back-off by 4dB.

The use of interleaving does not have any effect on the bit error rate performance of PC-OFDM systems. Interleaved PC-OFDM systems therefore possess the superior performance in AWGN channels compared to OFDM systems.

5. CONCLUSIONS

This chapter presents the use of data permutation in PC-OFDM to design systems with improved PAR statistics. PAR and BER is lower for PC-OFDM systems for high E_b / N_0. Use of interleaving will not effect the BER performance but reduces the PAR significantly. Theoretical expressions are presented to evaluate the CCDF of PC-OFDM and interleaved PC-OFDM signals. These formulae give simple means to evaluate the performance of different PC-OFDM systems with varying number of nonzero subcarriers, without performing extensive computer simulations.

REFERENCES

[1] L. J. Cimini, "Analysis and Simulation of a Digital Mobile Channel Using Orthogonal Frequency Division Multiplexing," *IEEE Trans. Comm.*, vol. COM-33, pp. 665-675, 1985.

[2] E. A. Jones, T. A. Wilkinson, and S. K. Barton, "Block Coding Scheme for Reduction of Peak-to-mean Envelope Power Ratio of Multicarrier Transmission Schemes," *IEE Elect. Letts.*, vol. 30, pp. 2098-2099, 1994.

[3] H. Muller and J. B. Huber, "OFDM with Reduced Peak-toAverage Power Ratio by Optimum Combination of Partial Transmit Sequences," *IEE Elect. Letts.*, vol. 33, pp. 368-369, 1997.

[4] P. K. Frenger and N. A. B. Sevensson, "Parallel Combinatory OFDM Signalling," *IEEE Trans. Commun.*, vol. 47, pp. 558-567, 1999.

[5] A. D. S. Jayalath and C. Tellambura, "Reducing the peak-to-average power ratio of an OFDM signal through bit or symbol interleaving," *IEE Elect. Letts..*, vol. 36, pp. 1161-1163, 2000.

[6] C. Tellambura, "Phase optimization criterion for reducing peak-to-average power ratio in OFDM," *Elect. Letts.*, vol. 34, pp. 169-170, 1998.

[7] H. Ochiai and H. Imai, "Performance of the deliberate clipping with adaptive symbol selection for strictly band-limited OFDM systems," *IEEE J. Select. Areas Commun.*, vol. 18, pp. 2270-2277, 2000.

[8] R. van-Nee, *OFDM wireless multimedia communications*. Boston, London: Artech House, 2000.

[9] H. E. Rowe, "Memoryless nonlinearities with Gaussian inputs: elementary results," *BELL Systems Tech. J.*, vol. 61, pp. 1519-1525, 1982.

Chapter 19

REDUCING PAR AND PICR OF AN OFDM SIGNAL

K. Sathananthan and C. Tellambura
Faculty of Information Technology, Monash University, Australia

Abstract: Orthogonal Frequency Division Multiplexing (OFDM), a promising technique to support high-bit-rate wireless systems, is limited by potentially large Peak-to-Average Power ratio (PAR) and Intercarrier Interference (ICI) due to frequency offset errors. Several techniques proposed in the literature treat these limitations as two separate problems. We show that a simple adaptive modulation scheme is effective in reducing the both PAR and ICI simultaneously. We propose the use of M-point Zero-padded Phase Shift Keying (M-ZPSK), which includes a signal point of zero amplitude, as a modulation scheme. The performance of this scheme is evaluated analytically and by simulation. We also introduce the Peak Interference-to-Carrier Ratio (PICR) to measure the resulting intercarrier interference (ICI) effects at the transmitter. We show that Partial Transmit Sequence (PTS) and Selected Mapping (SLM) approaches can reduce PICR. These schemes are analyzed theoretically and their performances are evaluated by simulation.

Key words: Orthogonal Frequency Division Multiplexing, Peak-to-Average Power ratio (PAR), Carrier Frequency Offset (CFO), Intercarrier Interference (ICI) and Peak Interference-to-Carrier Ratio (PICR)

1. INTRODUCTION

Orthogonal Frequency Division Multiplexing (OFDM) is an attractive technique to support high-data-rate wireless services. It has been accepted for several wireless LAN standards, as well as a number of mobile multimedia applications [1]. However, two major drawbacks of OFDM are its potentially large Peak-to-Average Power ratio (PAR) and Intercarrier Interference (ICI) caused by carrier frequency offset (CFO). Due to the large number of subcarriers, the OFDM signal has a large dynamic signal range with high PAR. These high signal peaks can lead to in-band distortion and spectral spreading in the presence of non-linear devices. This results in performance degradation and/or inefficiency in use of the non-linear devices [2-7]. The cause for CFO is the misalignment in carrier frequencies between the transmitter and receiver or by Doppler shift [8, 9]. Increased interest in OFDM for many wireless applications has motivated the search for techniques to overcome these impairments.

The key idea for the PAR reduction technique in [6, 7] is to expand the M-PSK constellation with an extra, zero amplitude point. Thus, some subcarriers have zero amplitude. This reduces the peak of the OFDM signal. Interestingly, these two approaches offer higher bandwidth efficiency than that of a normal OFDM system with increased complexity as non-modulated subcarrier positions are also used to carry information bits. However, the selection of subcarriers with zero amplitude is complex and inefficient when N is large. Therefore, we propose an M-ZPSK modulation scheme, which includes a signal point of zero amplitude. Thus, some subcarriers have zero amplitude. Note that both M-ZPSK and M-PSK have the same number of signal points. We now propose a simple adaptive M-ZPSK modulation scheme to reduce the both PAR and ICI simultaneously.

We also introduce Peak Interference-to-Carrier Ratio (PICR) to measure the ICI effects at the transmitter. Note that the PAR problem is at the transmitter where as ICI is at the receiver. Therefore, ICI effects can be quantified at the transmitter in order to easily device a common scheme to reduce both PAR and ICI simultaneously. Moreover, formulating ICI problem at the transmitter is advantageous to device a new and effective ICI reduction schemes. For instance, codes can be designed to reduce ICI. Interestingly, the definition of PICR formulates the ICI problem in parallel with PAR issue. Generating several statistically independent OFDM frames representing the same information sequence and selecting the one with the lowest PAR is a common approach in [4, 10]. This approach improves the statistics of the PAR of an OFDM signal with additional complexity. Motivated by the success of these approaches, we apply the PTS in [4, 10] and SLM in [4] approaches to reduce PICR.

The organization of this chapter is as follows: Section II presents the PAR and ICI problems in OFDM systems. Adaptive Modulation scheme is explained in Section III. We introduce PICR problem and application of SLM and PTS approach to reduce PICR in Section IV. Concluding remarks are presented in Section V.

2. OFDM ISSUES

2.1 OFDM Signaling

The complex baseband OFDM signal may be represented as

$$s(t) = \frac{1}{\sqrt{N}} \sum_{k=0}^{N-1} c_k e^{j2\pi k\Delta f t} \tag{1}$$

where $j^2 = -1$ and N is the total number of subcarriers. The frequency separation between any two adjacent subcarriers is $\Delta f = 1/T$ and T is the OFDM symbol duration. c_k is the data symbol for the k-th subcarrier. Each modulated symbol c_k is chosen from the set $F_M = \{\lambda_1, \lambda_2, .., \lambda_M\}$ of M distinct elements. The set F_M is called *signal constellation* of the M-ary modulation scheme. We shall refer to $\mathbf{c} = (c_0, c_1, .., c_{N-1})$ as a data frame or codeword, as appropriate and is a constellation symbols from an encoder.

2.2 PAR Problem

The PAR of the transmitted signal in (1) can be defined as

$$PAR(\mathrm{c}) = \frac{\max|s(t)|^2}{E[|s(t)|^2]} \tag{2}$$

where $E[\cdot]$ denotes the average. Therefore, the maximum PAR of a baseband signal can be expressed as

$$\max PAR(\mathbf{c}) = N \tag{3}$$

To more accurately approximate the true PAR, $s(t)$ has to be oversampled. In this study, we consider oversampling by a factor of four, as further increase in oversampling does not influence the PAR.

2.3 ICI Problem

We assume that $s(t)$ is transmitted on an additive white Gaussian noise channel, and so the received signal sample for the k-th subcarrier after Discrete Fourier Transform (DFT) demodulation can be written as

$$y_k = c_k S_0 + \sum_{l=0, l \neq k}^{N-1} c_l S_{l-k} + n_k \tag{4}$$

where n_k is a complex Gaussian noise sample (with its real and imaginary components being independent and identically distributed with variance σ^2) and c_k is one of the signal constellations. The second term in (4) is the ICI term attributable to the CFO. The sequence S_k (the ICI coefficients) depends on the CFO and is given by

$$S_k = \frac{\sin \pi(k+\varepsilon)}{N \sin \frac{\pi}{N}(k+\varepsilon)} \exp[j\pi(1-\tfrac{1}{N})(k+\varepsilon)] \tag{5}$$

where ε is the normalized frequency offset defined as a ratio between the frequency offset (which remains constant over each symbol period) and the subcarrier spacing. For a zero frequency offset, S_k reduces to the unit impulse sequence. The ICI term can be expressed as

$$I_k = \sum_{l=0, l \neq k}^{N-1} c_l S_{l-k} \tag{6}$$

Note that I_k is a function of both c_k and ε.

3. PAR REDUCTION BY ADAPTIVE MODULATION

3.1 M-ZPSK Modulation Scheme

The signal constellation of M-ZPSK can be expressed as

$$F_M = \left[0, \alpha, \alpha\xi, .., \alpha\xi^{M-1}\right] \tag{7}$$

where $\alpha = \sqrt{\dfrac{M}{M-1}}$ and $\xi = e^{j\frac{2\pi}{M-1}}$.

The average power of these signal points is one. Further, this new mapping does not require any redundant subcarriers to deliberately introduce zero amplitude symbols in an OFDM block. Thus, the bandwidth efficiency and the complexity of this mapping is the same as normal OFDM system, unlike in [6, 7]. However, one disadvantage of M-ZPSK over conventional

M-PSK is the degradation in signal-to-noise ratio (SNR) as signal points are little closer in M-ZPSK. The minimum Euclidean distance of M-ZPSK is

$$\min\left\{2\sqrt{\frac{M}{M-1}}\sin\left\{\frac{\pi}{M-1}\right\},\sqrt{\frac{M}{M-1}}\right\}$$

whereas that of M-PSK is

$$2\sin\left\{\frac{\pi}{M}\right\}.$$

For an example, the minimum Euclidean distance is 1.4142 for 4-PSK and 1.1547 for 4-ZPSK.

3.2 Adaptive Approach for PAR Reduction

The maximum PAR of a baseband OFDM signal with M-ZPSK is expressed as

$$\max PAR(\mathbf{c}) = N - L \tag{8}$$

where L is the number of zero amplitude signal points in the OFDM block. Note that L is random and it determines the PAR reduction capability of the M-ZPSK mapping. Thus, PAR reduction may not be significant for small L. If all the signal points in M-ZPSK have the same a priori probability, then, L is equal to N/M. For $N = 128$ and $M = 4$, the maximum PAR is 19.8 dB whereas that of conventional MPSK is 21 dB. However, the PAR reduction of 1.2 dB due to M-ZPSK mapping may not be realized as the maximum PAR occurs very rarely in practice.

To exploit the characteristics of the M-ZPSK mapping, we propose an adaptive scheme to reduce PAR for any random data. In this scheme, one out of M bit patterns of $\log_2 M$ bits can be mapped to a signal constellation of zero amplitude. This leads to M independent OFDM frames representing the same information. Therefore, the OFDM frame with low PAR (say PAR_{low}) can be selected for transmission. The block diagram for this scheme and constellation mapping for $M = 4$ are shown in *Figure 1* and *Figure 2* respectively.

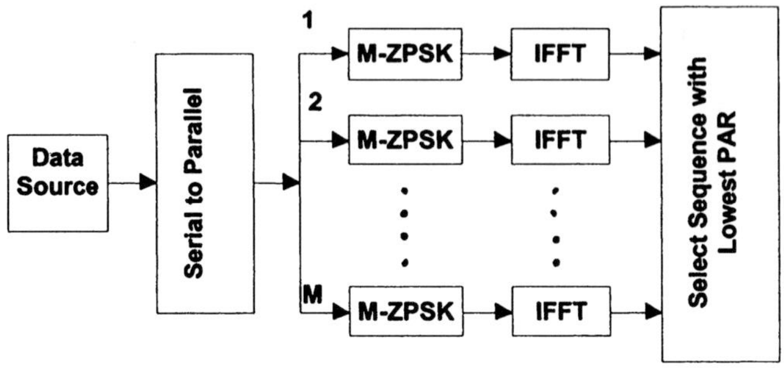

Figure 1. Block Diagram for Adaptive OFDM Scheme

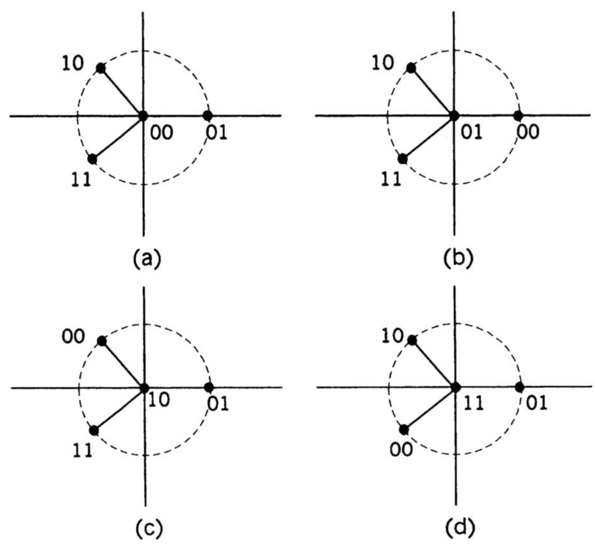

Figure 2. Mapping Signal Constellations for Adaptive OFDM for $M = 4$

To recover the data, the receiver has to know which mapping has actually been used in the transmitter. This can be achieved by transmitting $\log_2 M$ bits as Side Information (SI). As these bits are most important, they should be protected by channel coding. The proposed scheme requires M number of data mapping blocks and IFFT blocks in the transmitter. Moreover, this approach applies to an arbitrary number of subcarriers. Note that this scheme does not require any multiplication and complex optimization as in [4]. Mapping of bits to constellation symbols is a rather simple operation comparing with multiplication.

The reduction capability of the proposed scheme is restricted by modulation scheme as M determines the number of independent OFDM frames representing the same information. Yet, this is not an issue as high peaks occur infrequently and it is desirable to keep the complexity minimum. Another important advantage of this scheme is reduced sensitivity to frequency offset errors. We will elaborate this in the sequel.

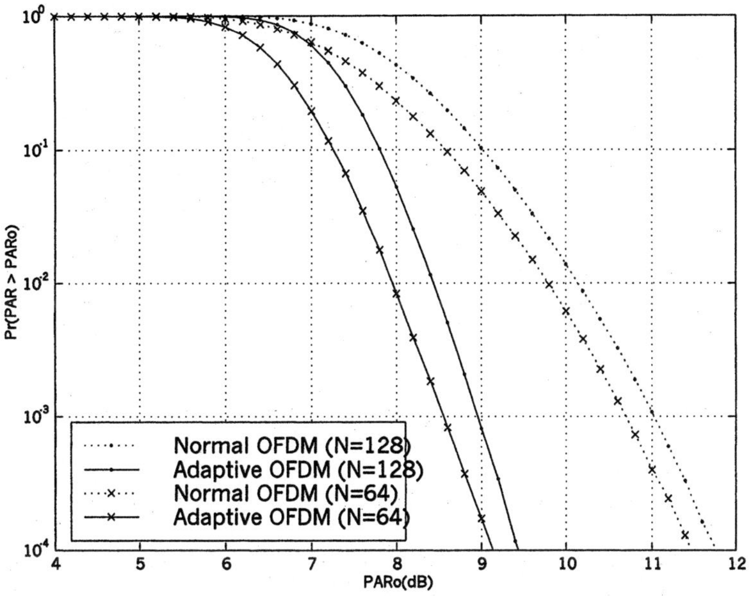

Figure 3. CCDF of the PAR of an OFDM Signal for N=128 and M=4.

Figure 1-3 shows the complementary cumulative density function (CCDF) of the PAR of an OFDM signal for $N = 64$ and $N = 128$ with $M = 4$. For $N = 128$, the PAR of an adaptive 4-ZPSK scheme exceeds 9 dB for only 1 out of 10^3 whereas that of conventional scheme is 1 out of 10.

3.2.1 Out of Band Radiation

OFDM signal is subject to various hardware non-linearities in both the transmitter and receiver. Hardware non-linearities not only affect the performance of an OFDM system, but also may affect the system performance of an adjacent channel because of generated side lobes [11]. Out of band power of OFDM signals increases, when amplified with nonlinear power amplifiers operating at lower back-offs. The high PAR of OFDM requires high back-offs at the amplifiers. Power Spectral Density

(PSD) of the OFDM signal and proposed adaptive OFDM signal is estimated via simulations. The PSD is estimated using Welch's averaged periodogram method with a Hanning window.

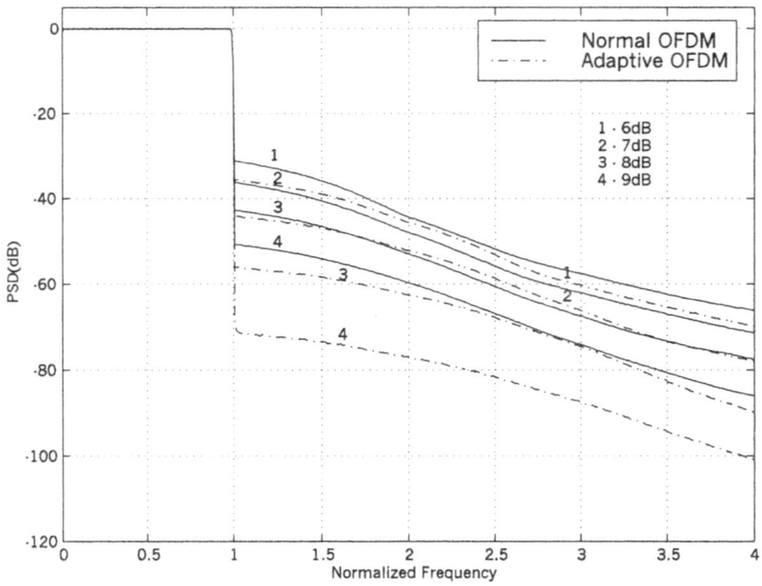

Figure 4. Spectrum of an OFDM Signal after passing through a SSPA

Figure 4 depicts the PSD in the presence of solid state power amplifier (SSPA) non-linear device for $N = 128$. Curves 1 to 4 represent the different input back-off values of the amplifier. For example, for a back-off of 8 dB, to provide -50 dB adjacent channel separation requires a channel spacing of about two times the symbol rate for normal OFDM, with adaptive OFDM the channel spacing is nearly the symbol rate. Thus, the adaptive OFDM provides a spectrally efficient solution for the adjacent channel interference problem, which can arise in portable cellular applications. The performances in *Figure 1-4* reveals that depending on the permissible out of band radiation level, the amplifier back-off can be reduced by 2 to 3 dB when adaptive OFDM is used instead of conventional OFDM.

3.2.2 ICI Reduction

In M-ZPSK, we have a signal point of zero amplitude. Thus, some terms in the summation in (6) vanishes. Therefore, the adaptive M-ZPSK scheme is less sensitive to frequency offset errors than conventional schemes. *Figure*

5 shows the BER of an adaptive 4-ZPSK modulated OFDM signal with $N = 128$ and $\varepsilon = 0.1$. The adaptive scheme offers an SNR gain of 2 dB over conventional scheme at the BER of 10^{-3} in AWGN channel.

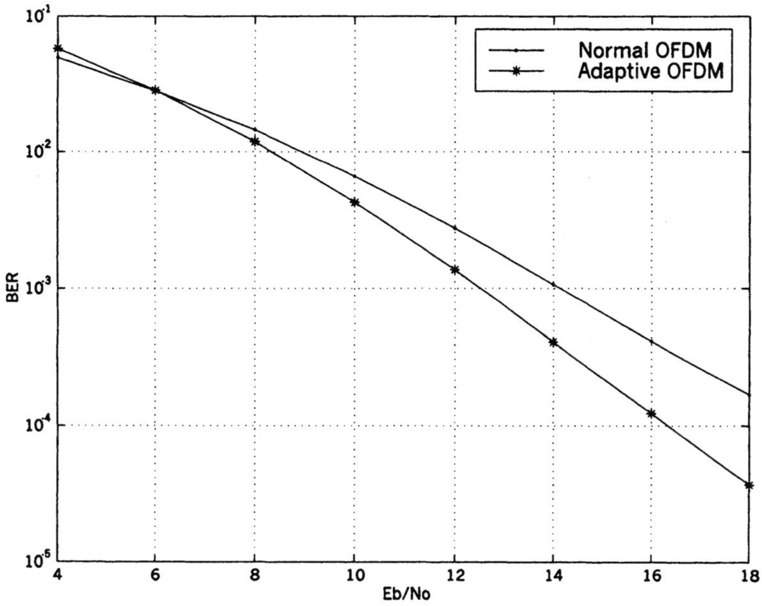

Figure 5. BER of the Adaptive OFDM Scheme in AWGN channel for $M = 4$ and $\varepsilon = 0.1$

Interestingly, simplified adaptive M-ZPSK mapping could be used for ICI reduction if PAR reduction is not taken into account. In this case, the frequency of the bit pattern of $\log_2 M$ bits in an input data frame is counted. The most likely bit pattern is mapped to a signal constellation of zero amplitude. This increases the vanishing terms in the summation in (6) and thus reduces the ICI effects. Therefore, we need only one mapping and one IFFT calculation, as in the conventional system. A simple counter circuit can be used to select a mapping scheme based on the input data sequence. However, transmission of side information is necessary as in PAR reduction criteria. Note that this approach will not assure the PAR reduction but rather simple approach to reduce ICI effects.

4. REDUCING PICR BY PTS AND SLM SCHEMES

We define the Peak Interference-to-Carrier Ratio (PICR) as

$$PICR(\mathbf{c},\varepsilon) = \max_{0 \le k \le N-1} \left\{ \frac{|I_k|^2}{|S_0 c_k|^2} \right\}. \tag{9}$$

Note that the PICR is a function of both $\mathbf{c}$ and ε. PICR is the maximum interference-to-signal ratio for any subcarrier. In other words, it specifies the worst-case ICI on any subcarrier. To reduce ICI effects, (9) should be minimized and is zero for ICI-free channels. Interestingly, our definition in (9) is similar to PAR issue in OFDM [2-7]. However, the PICR problem differs from the PAR issue in several ways:

- ICI occurs at the receiver side, whereas high PAR values affect the transmitter
- Exact computation of PAR requires oversampling, whereas $\max|I_k|$ is obtained from N samples.
- As the transmitter does not know ε a priori, PICR can be computed only on the basis of a worst-case value, $\varepsilon_{wc} > 0$. Therefore, the performance of a PICR reduction scheme should hold for any $|\varepsilon| < \varepsilon_{wc}$.

Figure 6 shows the CCDF of the PICR as a function of N for $\varepsilon = 0.1$. The subcarriers are modulated with quadrature phase shift keying (QPSK). For $N = 128$, the PICR exceeds –5.5 dB for only 1 out 10^4 of all OFDM blocks. Therefore, the PICR can be considered as a random variable and is dependent on data. Naturally, one may apply PAR reduction techniques, which exploits the statistical properties of an OFDM signal, to reduce PICR. As PTS and SLM schemes can reduce PAR, we investigate them to reduce PICR.

In the SLM scheme, several independent OFDM symbols representing the same information are generated by multiplying the information sequence by a set of fixed vectors. Then, the OFDM symbol with lowest PICR is selected for transmission. The receiver must know which multiplying vector has been used. A pointer to this sequence is transmitted as side information. In the PTS scheme, subcarriers are partitioned into blocks and each block is multiplied by a constant phase factor. These phase factors are optimized to minimize the PICR. Optimal phase factors are sent to the receiver as side information.

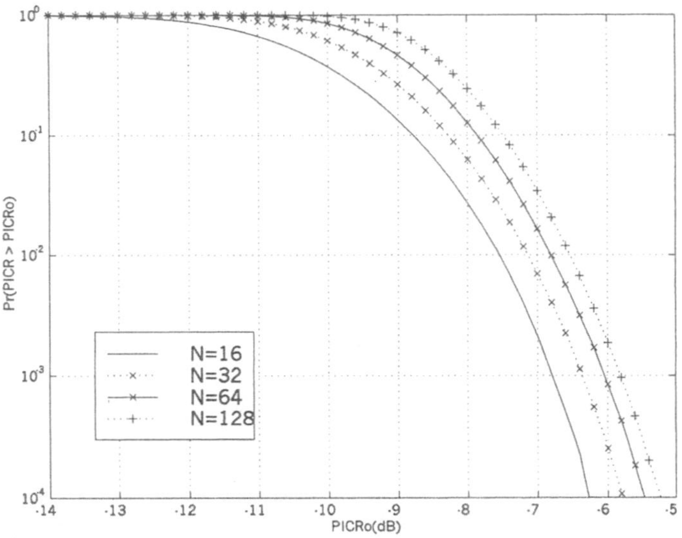

Figure 6. CCDF of PICR of Normal OFDM System with $\varepsilon = 0.1$

4.1 Selected Mapping (SLM)

U statistically independent alternative transmit sequences $\mathbf{a}^{(u)}$ represent the same information. The sequence with lowest PICR is selected for transmission. To generate $\mathbf{a}^{(u)}$, we define U distinct fixed vectors $\mathbf{P}^{(u)} = [P_0^{(u)},..,P_{N-1}^{(u)}]$ with, $P_v^{(u)} = e^{j\phi_v^{(u)}}$, $\phi_v^{(u)} \in [0,2\pi]$, $0 \le v \le N-1$ and $1 \le u \le U$. Then, each modulated symbol $\mathbf{c} = [c_0, c_1,.., c_{N-1}]$ is multiplied carrierwise with the U vectors $\mathbf{P}^{(u)}$, resulting in a set of U different modulated symbols $\mathbf{c}^{(u)}$ with components

$$c_v^{(u)} = c_v P_v^{(u)} \tag{10}$$

For simple implementation, we select $P_v^{(u)} \in [\pm 1, \pm j]$ for $0 \le v \le N-1$ and $1 \le u \le U$. Using Eqs. (6) and (10), the resulting ICI on the *k*-th subcarrier can be expressed as

$$I_{k,SLM} = \sum_{l=0, l \ne k}^{N-1} P_l^{(u)} c_l S_{l-k} \tag{11}$$

which is a function of the weighting sequence $\mathbf{P}^{(u)}$. Finally, the optimal PICR can be found as

$$PICR_{optimul} = \min_{\mathbf{P}^{(1)},..,\mathbf{P}^{(U)}} \left[\frac{\max\limits_{0 \le k \le N-1} \left| I_{k,SLM} \right|^2}{\left| S_0 c_k \right|} \right]. \tag{12}$$

The computation of PICR at the transmitter requires ε. However, a worst-case ε can be assumed for the computation. Simulation results show that optimized phase factors for ε_{wc} are effective for another ε provided $|\varepsilon| < \varepsilon_{wc}$. That is, any mismatch between the actual CFO and the worst-case CFO can be handled under this condition.

4.2 Partial Transmit Sequences (PTS)

In PTS, the input data block is partitioned into disjoint subblocks or clusters, which are combined to minimize the peaks. We partition the data frame $\mathbf{c}$ into M disjoint subblocks, represented by the vectors $\{\mathbf{c}_m, m = 1,2,..,M\}$, such $\mathbf{c} = [\mathbf{c}_1, \mathbf{c}_2,..,\mathbf{c}_M]$. It is assumed that each subblock consists of a contiguous set of subcarriers and the subblocks are of equal size. Then, each subblock is zero padded to make its length N and is multiplied by a weighting factor $b_m (m = 1,2,..,M)$. Thus, the ICI on the k-th subcarrier can be expressed as

$$I_{k,PTS} = \sum_{m=1}^{M} \sum_{l=0,l \ne k}^{N-1} b_m c_l^m S_{l-k} \tag{13}$$

where c_l^m is the data symbol in the newly formed m-th subblock. We can write (13) as

$$I_{k,PTS} = \sum_{m=1}^{M} b_m I_k^m \tag{14}$$

where I_k^m is the interference on k-th subcarrier of block m. Thus, the total ICI is the weighted sum of ICI from each subblock and the total ICI and PICR can be reduced by optimizing the phase sequence $\mathbf{b} = [b_1, b_2,..,b_M]$. A drawback in the PTS approach is the complexity of the optimization of phase factors. To reduce of this complexity, we only consider binary phase factors (i.e., $b_m = \pm 1$). Without loss of generality, we can set b_1 and observe that there are $(M-1)$ binary variables to be optimized. Finally, the optimal PICR can be found as

$$PICR_{optimal} = \min_{b_1,..,b_M} \left[\frac{\max\limits_{0 \le k \le N-1} \left| I_{k,PTS} \right|^2}{\left| S_0 c_k \right|^2} \right] \tag{15}$$

The worst-case ε can be assumed for the computation of PICR, as in SLM.

4.3 Simulation Results

The simulation results were obtained for an OFDM system with $N = 128$. The subcarriers are modulated with QPSK and an AWGN channel is assumed throughout this study. Further, a worst-case ε of 0.1 is assumed.

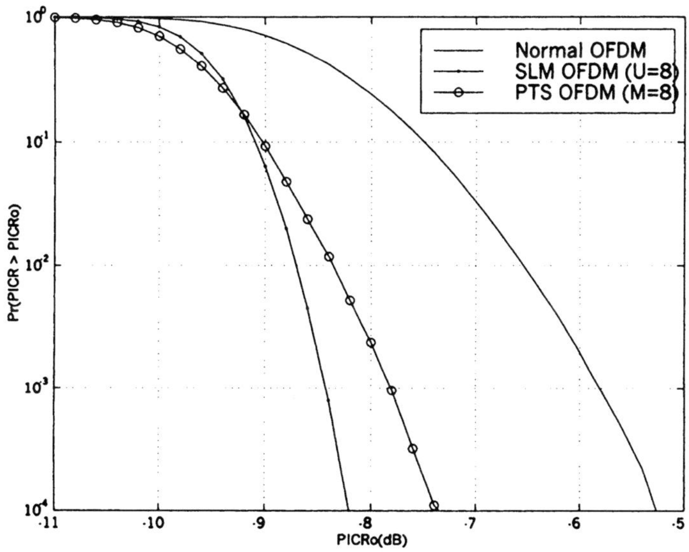

Figure 7. CCDF of PICR of SLM and PTS OFDM $\varepsilon = 0.1$

Figure 7 shows the CCDF of PICR per OFDM block for SLM and PTS for $\varepsilon = 0.1$. In SLM approach with $U = 8$, only 1 out 10^4 of all OFDM blocks exceeds the PICR of -8.5 dB whereas in normal OFDM, 1 out 10^4 of all OFDM blocks exceeds the PICR of -5.5 dB. That amounts to a 3.0 dB reduction in PICR. In the PTS approach, the PICR exceeds –7.5 dB for only 1 out 10^4 of all OFDM blocks whereas that of normal OFDM is for only 1 out 10. There is 2 dB reductions in PICR over normal OFDM with $M = 8$.

Moreover, PICR reduction increases with increasing U and M. However, the computational complexity also increases with U and M. Thus, the performance can be traded off against complexity.

Figure 8 shows the CCDF of PICR per OFDM block as a function of ε for both PTS and SLM. The PICR can be computed by assuming worst-case CFO and the performance of SLM and PTS will not be affected by CFO less than the worst-case CFO.

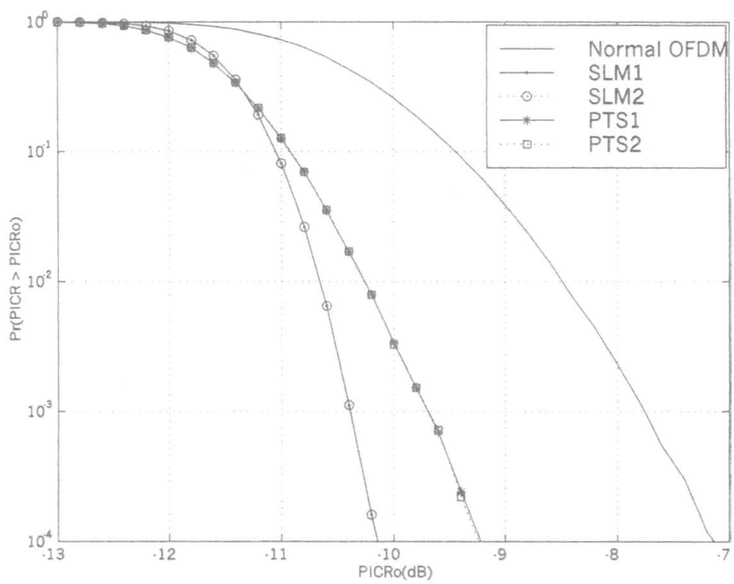

Figure 8. CCDF of PICR of an OFDM System (SLM1 & PTS1: $\varepsilon = 0.08$ and $\varepsilon_{wc} = 0.08$; SLM2 & PTS2: $\varepsilon = 0.08$ and $\varepsilon_{wc} = 0.1$)

5. SUMMARY

This chapter proposed an adaptive M-ZPSK to reduce both PAR and ICI of an OFDM signal simultaneously. The PAR improvement of the proposed scheme is comparable with other well-known approaches [4, 10] with almost same or less complexity. Simulation and analytical results show that adaptive 4-ZPSK modulated OFDM system with 128 subcarriers improves the PAR by 2.5 dB at the expense of a 2 dB loss in SNR over conventional system. However, in a channel with normalized frequency offset of 0.1, it offers a 2 dB gain in SNR at BER of 10^{-3} over conventional scheme. Moreover, for a back-off of 8 dB, to provide -50 dB adjacent channel separation requires a channel spacing of about two times the symbol rate for normal OFDM, with adaptive OFDM the channel spacing is nearly the symbol rate, which is a 50% saving in bandwidth. The price for these benefits is increased complexity over conventional OFDM systems.

This chapter also introduced PICR to measure ICI effects at the transmitter. The definition of PICR is analogous to that of PAR. Consequently, PAR reduction schemes can be applied to reduce PICR. We investigated the PTS and SLM methods to reduce PICR. They improve the PICR statistics of an OFDM signal at the expense of additional complexity, but with little loss in efficiency. For an OFDM system with $N = 128$ and $\varepsilon_{wc} = 0.1$, PTS with $M = 8$ reduces PICR by 2 dB whereas SLM with $U = 8$ reduces PICR by 3 dB. Moreover, both schemes work independent of ε, provided $|\varepsilon| < \varepsilon_{wc}$.

REFERENCES

[1] R. V. Nee and R. Prasad, *OFDM for wireless multimedia communications*, Artech House Publishers, 2000.

[2] E. A. Jones, T. A. Wilkinson, and S. K. Barton, "Block Coding Scheme for Reduction of Peak-to-mean Envelope Power Ratio of Multicarrier Transmission Schemes," *Elect. Letts.*, pp. 2098-2099, 1994.

[3] X. Li and L. J. Cimini, "Effects of clipping and filtering on the performance of OFDM," *IEEE Commun. Letts.*, vol. 2, pp. 131-133, 1998.

[4] S. H. Muller, R. W. Bauml, R. F. H. Fischer, and J. B. Huber, "OFDM with Reduced Peak-to-Average Power Ratio by Multiple Signal Representation," *Annals of Telecommunications*, vol. 52, pp. 58-67, Feb. 1997.

[5] K. G. Paterson and V. Tarokh, "On the Existence and Construction of Good Codes with Low Peak-to-Average Power Ratios," *IEEE Trans. Inform. Theory*, vol. 46, pp. 1974-1987, Sep. 2000.

[6] P. K. Frenger and N. A. B. Sevensson, "Parallel Combinatory OFDM Signalling," *IEEE Trans. Commun.*, vol. 47, pp. 558-567, Apr. 1999.

[7] A. Sumasu, T. Ue, M. Usesugi, O. Kato, and K. Homma, "A method to reduce the peak power with signal space expansion (ESPAR) for OFDM system," in *IEEE Vehicular Technology Conference*, pp. 405-409, 2000.

[8] P. Moose, "A Technique for Orthogonal Frequency Division Multiplexing Frequency Offset Correction," *IEEE Trans. Commun.*, vol. 42, pp. 2908-2914, 1994.

[9] J. Armstrong, "Analysis of new and existing methods of reducing intercarrier interference due to carrier frequency offset in OFDM," *IEEE Trans. Commun.*, vol. 47, pp. 365-369, Mar. 1999.

[10] L. J. Cimini and N. R. Sollenberger, "Peak-to-Average Power Ratio Reduction of an OFDM Signal Using Partial Transmit Sequences," *IEEE Commun. Letts.*, vol. 4, pp. 511-515, 1999.

[11] S. Merchan, A. G. Armada, and J. L. Garcia, "OFDM Performance in Amplifier Nonlinearity," *IEEE Trans. Broad.*, vol. 44, pp. 106-114, 1998.

Chapter 20

ITERATIVE JOINT EQUALIZATION AND DECODING BASED ON SOFT CHOLESKY EQUALIZATION FOR GENERAL COMPLEX VALUED MODULATION SYMBOLS

Jochem Egle and Jürgen Lindner
Department of Information Technology, University of Ulm, Germany

Abstract: A new iterative equalizer based on the classical block decision feedback equalizer is introduced. The performance of this iterative version benefits from the usage of a soft decision function, which is based on noise whitening and consecutive matched filtering. The derivation of the decision function is very general and, therefore, applicable to arbitrary complex valued modulation symbols. After adding the possibility of processing extrinsic information to this decision function, the equalizer is embedded in an iterative equalization/decoding scheme. To demonstrate the performance of the scheme simulation results are presented for frequency selective channels using BPSK and 16 QAM.

Key Words: Equalization, Decoding, Soft Decision

1. INTRODUCTION

In recent years mobile communication has seen a tremendous development. The growth rate of number of subscribers for mobile communication systems is enormous and mobile multimedia communication demanding very high data rate is on the horizon. Since the available spectrum allocation for mobile communication systems and the transmit power are limited, suitable methods have to be found, which use the available bandwidth efficiently and require as little transmit power as possible. To meet these requirements, many modern commutation systems use non-binary modulation alphabets, e. g. , M-ary phase shift keying (PSK) or M-ary quadrature amplitude modulation (QAM) schemes in combination with forward error correcting codes.

In the following, we introduce an algorithm which is capable to detect a coded signal which uses an arbitrary complex valued modulation alphabet and is subject to a high level of inter symbol interference (ISI) originating from a frequency selective channel. A detection scheme with moderate complexity and often near optimum performance is based on recurrent neural networks (RNN). It was introduced first for systems without coding in [13] for the case of BPSK and extended to general complex valued modulation alphabets in [11]. In [9] and [10] it is generalized to systems with coding. Other approaches leading to equalizer structures closely related to the one derived here are explained in [1] and [5]. In both cases linear filtering in combination with iterative soft decision feedback is used. The difference between both is the filter design criterion. The first one uses the unbiased minimum variance criterion whereas the second one the minimum mean square criterion. A first simplified version of the equalizer structure investigated in the following is explained in [3]. There, in a multi-user context, for the first time the ideas of multiple iterations and soft feedback are applied to the classical cholesky based block decision feedback equalizer (CBDFE).

Using this as a basis, we drop some simplifying assumptions and generalize the results previously obtained, delivering a set of equalizers which differ in complexity and performance. Furthermore, we test the suitability of this new equalizer to be used in a "Turbo" equalization scheme similar to the structure introduced first in [2].

After introducing a vector valued model of a coded transmission over a frequency selective channel based on [6], we describe the classical CBDFE. Starting from this basis, we derive a new iterative equalizer, i. e. , the iterative soft Cholesky block decision feedback equalizer (SCE), for an uncoded transmission. Subsequently, we add the possibility of processing extrinsic information to the SCE and use this modified version in an iterative coding and equalization scheme. In the simulation part we compare the performance of different versions of the SCE against each other, to the matched filter bound (MFB) and to an RNN type equalizer, see e. g. [11] for coded and uncoded transmission.

For mathematical notation we use double underlined letters to denote matrices $\underline{\underline{M}}$, whereas vectors $\underline{v}$ are denoted by single underlined letters. Furthermore, $(\cdot)^T$ denotes the transposed of a matrix, $(\cdot)^H$ the hermitian of a matrix, i. e. , the transposed conjugate complex of a matrix, and $(\cdot)^{-1}$ the inverse of a matrix.

2. SYSTEM MODEL

Figure 2 depicts a continuous-time model of a coded single carrier transmission in the equivalent low-pass domain. To indicate equivalent low-pass signals we use subscript T. A sequence of statistical independent bits $\underline{q}$ is encoded using an punctured terminated convolutional code. The code is deduced from a rate $1/n$ mother code using a puncturing matrix $\underline{\underline{P}}$. After interleaving the encoded bit sequence using a pseudo random interleaver matrix $\underline{\underline{\Pi}}$ the modulation specific coding COD$_{\text{MOD}}$ groups the interleaved binary codeword sequence $\underline{c}$ into blocks of $\log_2(M)$ bits. Each of those blocks is determining one complex transmit symbol $x(k) \in \mathcal{A} = \{a_1, a_2, \ldots, a_M\}$. In "Guard", periodically after N transmit symbols, a guard period of known symbols is inserted such that the receive sequence may be separated into independent blocks in the receiver. For simplicity, we assume the insertion of zeros . After parallel/serial (P/S) conversion the transmit symbols $x_g(k)$ are modulated using the basic waveform $e_T(t)$ resulting in the transmit signal $s_T(t)$. Subsequently, the modulated

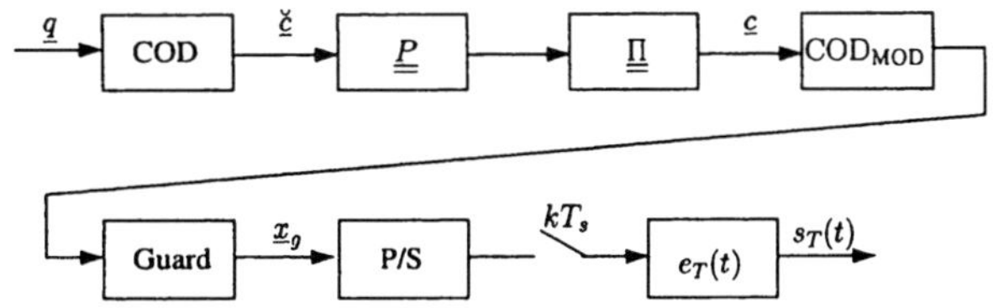

Figure 1. Transmit side of a coded single carrier transmission scheme

signal $s_T(t)$ is transmitted over a frequency selective additive white Gaussian noise channel having impulse response $h_T(t)$. In the receiver the symbol time sampled output sequence of the channel matched filter $h_{\text{CMF}_T}(t)$

$$h_{\text{CMF}_T}(t) = \frac{1}{2} e_T^*(-t) * h_T^*(-t) \tag{1}$$

is fed to an equalization/decoding algorithm. Introducing the discrete-time

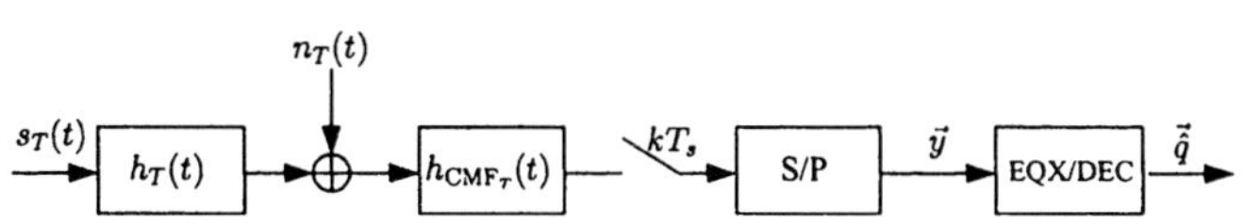

Figure 2. Transmit channel together with a receiver

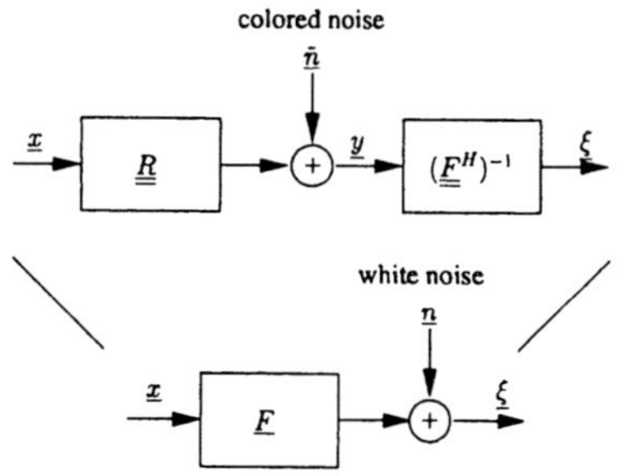

Figure 3. Equivalent vector valued transmission scheme

auto correlation r(k) and the sampled colored noise $\tilde{n}(k)$

$$\begin{aligned} r(k) &= \frac{1}{4} h_{\mathrm{CMF}_T}(t) * h_T(t) * e_T(t)|_{t=kT_s} \\ \tilde{n}(k) &= \frac{1}{2} h_{\mathrm{CMF}_T}(t) * n_T(t)|_{t=kT_s} \end{aligned} \tag{2}$$

permits to reformulate the relation between the vector of transmit symbols $\underline{x} = (x(1), \ldots, x(N))^T$ and the sampled output of the matched filter (without guard time) $\underline{y} = (y(1), \ldots, y(N))^T$ in compact matrix vector notation

$$\begin{aligned} &\underline{y} = \underline{\underline{R}}\, \underline{x} + \underline{\tilde{n}} \text{ with} \\ &\underline{\underline{R}} = \underline{\underline{[r_{ij}]}} \text{ and } r_{ij} = r(i-j)\ . \end{aligned} \tag{3}$$

Note, in order to get a concise notation the block indices are omitted.

3. EQUALIZATION SCHEME

Commencing with a conventional Cholesky based block decision feedback equalizer (CBDFE), which may be viewed as the block processing counterpart of the ordinary decision feedback equalizer (DFE) based on finite impulse response (FIR)-filtering, we derive an iterative block decision feedback equalizer. Retaining the filter matrices of the CBDFE, the new scheme overcomes the unavoidable SNR-loss of a conventional CBDFE using a novel soft decision function and multiple iterations. Therefore, we term this new scheme iterative soft Cholesky block decision feedback equalizer (SCE).

3.1 Cholesky Block Decision Feedback Equalizer

$\underline{\underline{R}}$, as a correlation matrix, is hermitian and positive semi-definite. Taking into account, that a positive semi-definite matrix becomes positive definite by adding a small positive value, e. g. , the noise variance, on the main diagonal $\underline{\underline{R}}$ can be treated as positive definite in the following. Then, $\underline{\underline{R}}$ can be factorized in a product of an upper triangular matrix $\underline{\underline{F}}$ and the hermitian of this matrix using the Cholesky decomposition:

$$\underline{\underline{R}} = \underline{\underline{F}}^H \underline{\underline{F}} \tag{4}$$

The CBDFE uses $(\underline{\underline{F}}^H)^{-1}$ as feed-forward filter. The combination of the channel matrix $\underline{\underline{R}}$ and the feed-forward filter $(\underline{\underline{F}}^H)^{-1}$ results in the simplified model shown in Figure 3 and described by

$$\underline{\xi} = (\underline{\underline{F}}^H)^{-1}(\underline{\underline{R}}\,\underline{x} + \underline{\tilde{n}}) = (\underline{\underline{F}}^H)^{-1}(\underline{\underline{F}}^H \underline{\underline{F}}\,\underline{x} + \underline{\underline{F}}^H \underline{n}) = \underline{\underline{F}}\,\underline{x} + \underline{n} \tag{5}$$

Using dec$(\cdot)$ as a two-dimensional decision device matched to the modulation alphabet $\mathcal{A}$ and exploiting that $\underline{\underline{F}}$ is an upper triangular matrix the CBDFE sequentially detects the block of transmitted symbols starting from the end of the block

$$\begin{aligned} \hat{x}_N &= \operatorname{dec}\left(\frac{1}{f_{NN}}\xi_N\right) \\ \hat{x}_{N-1} &= \operatorname{dec}\left(\frac{1}{f_{N-1N-1}}(\xi_{N-1} - f_{N-1N}\hat{x}_N)\right) \\ &\vdots \qquad\qquad \vdots \\ \hat{x}_1 &= \operatorname{dec}\left(\frac{1}{f_{11}}(\xi_1 - \sum_{i=2}^{N} f_{1i}\hat{x}_i)\right) \quad . \end{aligned} \tag{6}$$

3.2 Iterative Soft Block Decision Feedback Equalizer

The soft decision function of the SCE is based on the optimum detector for colored Gaussian noise. In each iteration of the equalizer the covariance matrix of the interference as well as the transmit symbol vector are estimated.

3.2.1 Real valued matrix notation

Below, we introduce a notation which describes the quadrature components of a baseband symbol not in terms of complex numbers but by a vector of length two. This representation enables us to handle general complex Gaussian random vectors, like they are required for the derivation of the SCE and which are not be described correctly in terms of a complex Gaussian distribution, e. g., a random vector consisting of complex elements with non-zero covariance between the quadrature components. A deeper insight into the properties of such random vectors is given in [7]. In this notation a complex vectors $\underline{v}$ of length a is represented by real vectors $\underline{\mathsf{v}}$ of length $2a$ and a complex matrix $\underline{\underline{M}}$ of dimension $a \times b$ by real matrices $\underline{\underline{\mathsf{M}}}$ of dimension $2a \times 2b$. The transformation rule is shown in Equation (7). $\otimes$ denotes the Kronecker product.

$$\begin{aligned}
\underline{\mathsf{v}} &= \left(\Re(v_1) \quad \Im(v_1) \quad \ldots \quad \Re(v_a) \quad \Im(v_a)\right)^T \\
&= \Re(\underline{v}) \otimes \begin{pmatrix} 1 \\ 0 \end{pmatrix} + \Im(\underline{v}) \otimes \begin{pmatrix} 0 \\ 1 \end{pmatrix} \\
\underline{\underline{\mathsf{M}}} &= \begin{pmatrix} \Re(m_{11}) & -\Im(m_{11}) & \ldots & -\Im(m_{1b}) \\ \Im(m_{11}) & \Re(m_{11}) & \ldots & \Re(m_{1b}) \\ \vdots & \vdots & & \vdots \\ \Im(m_{a1}) & \Re(m_{a1}) & \ldots & \Re(m_{ab}) \end{pmatrix} \\
&= \Re(\underline{\underline{M}}) \otimes \begin{pmatrix} 1 & 0 \\ 0 & 1 \end{pmatrix} + \Im(\underline{\underline{M}}) \otimes \begin{pmatrix} 0 & -1 \\ 1 & 0 \end{pmatrix}
\end{aligned} \tag{7}$$

In this notation a complex symbol corresponds to a real vector of length 2, e. g. , the modulation alphabet is given by $\mathcal{A} = \{\underline{\mathsf{a}}_1, \ldots, \underline{\mathsf{a}}_M\}$ with $\underline{\mathsf{a}}_j = (\Re(a_j), \Im(a_j))^T$.

3.2.2 SCE – iteration scheme

The SCE replaces the hard decision function $\mathrm{dec}(\cdot)$ by a soft decision function $\mathrm{dec}_{\mathrm{soft}}(\cdot)$ which is designed under the constraint of minimizing the mean square error (MSE) of the decision. Furthermore, the SCE uses multiple iterations, each iteration starting from the end of the block, during which previous estimates of interfering symbols are fed back. A model of the SCE when calculating an estimate $\tilde{\underline{\mathsf{x}}}_l^{(\eta)}$ of the l-th symbol $\underline{\mathsf{x}}_l$ in the η-th iteration is shown in Figure 4 and described by

$$\begin{aligned}
\tilde{\underline{\mathsf{x}}}_l^{(\eta)} &= \mathrm{dec}_{\mathrm{soft}}(\underline{\xi}^{(\eta,l)} \underline{\underline{\Phi}}_{\mathsf{w}\,\mathsf{w}}^{(\eta,l)}) \quad \text{with} \quad \underline{\xi}^{(\eta,l)} = \underline{\xi} - \underline{\underline{\mathsf{F}}}_{\backslash l} \tilde{\underline{\mathsf{x}}}^{(\eta,l)} \\
\text{and} \quad \tilde{\underline{\mathsf{x}}}^{(\eta,l)} &= [\tilde{\underline{\mathsf{x}}}_1^{(\eta-1)^T}, \ldots, \tilde{\underline{\mathsf{x}}}_l^{(\eta-1)^T}, \tilde{\underline{\mathsf{x}}}_{l+1}^{(\eta)\,T}, \ldots, \tilde{\underline{\mathsf{x}}}_N^{(\eta)^T}]^T \; .
\end{aligned} \tag{8}$$

$\underline{\underline{\mathsf{F}}}_{\backslash l}$ denotes a matrix derived from $\underline{\underline{\mathsf{F}}}$ by setting the $(2l-1)$-th and the $2l$-th column to the zero vector. $\tilde{\underline{\mathsf{x}}}^{(\eta,l)}$ contains previous soft decisions of interfering symbols. The matrix $\underline{\underline{\Phi}}_{\mathsf{w}\,\mathsf{w}}^{(\eta,l)}$ is the covariance matrix of the noise plus interference signal. Initially, no estimates of the transmit symbols are available. Therefore, we initialize the feedback vector $\tilde{\underline{\mathsf{x}}}^{(1,N)}$ by the unconditioned mean

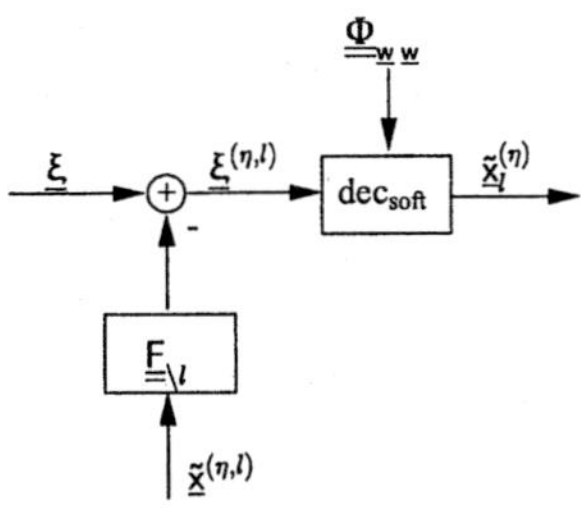

Figure 4. SCE updating estimate of symbol $\underline{x}_l$ in iteration η

of the transmit vector, $E\{\underline{x}\}$, which corresponds under the assumption of unbiased and equiprobable transmit symbols to the all-zero vector. Furthermore, let $\underline{\underline{f}}_l$ be the sub-matrix of $\underline{\underline{F}}$ which contains the $(2l-1)$-th and $2l$-th column of $\underline{\underline{F}}$, i. e. , the sub-matrix reflects the impact of the l-th transmit symbol $\underline{x}_l$ on the receive vector $\underline{\xi}$. Then, the desired end state is reached for $\tilde{\underline{x}}^{(\eta,l)} = \underline{x}$, i. e. , all interference has been subtracted and therefore $\underline{\xi}^{(\eta,l)}$ equals

$$\underline{\xi}^{(\eta,l)} = \underline{\underline{f}}_l \underline{x}_l + \underline{n} \ . \tag{9}$$

In this case $\underline{\xi}^{(\eta,l)}$ is only dependent on the transmit symbol $\underline{x}_l$.

3.2.3 Decision function

First, in [12] a decision rule aiming to minimize the mean square error (MSE) of the decision was proposed and further investigated in [4]. In [11] a similar idea was applied to determine the decision function of a RNN. Then in [3] the ideas mentioned above were applied to the CBDFE to overcome the DFE inherent SNR-loss. In the following, we generalize the results obtained in [3] resulting in a set of decision functions. In detail, we derive a decision function permitting to take decision based on all entries of $\underline{\underline{f}}_l$, whereas the decision function of a conventional CBDFE benefits only from entry on the main diagonal of $\underline{\underline{F}}$. A model of the decision of symbol $\underline{x}_l$ is shown in Figure 5. Multiplying the transmit symbol $\underline{x}_l$ with $\underline{\underline{f}}_l$ spreads its energy over the receive vector $\underline{\xi}^{(\eta,l)}$.

$$\underline{\xi}^{(\eta,l)} = \underline{\underline{f}}_l \underline{x}_l + \underline{n} + \underline{i}^{(\eta,l)} \tag{10}$$

Two parts disturb the reception. The first one, $\underline{n}$, is caused by additive white Gaussian noise which is assumed to be independent of the transmit signal, whereas the second part, $\underline{i}^{(\eta,l)}$, comprises the remaining interference of other symbols. For the sake of computational tractability, we treat the remaining interference $\underline{i}^{(\eta,l)}$

$$\underline{i}^{(\eta,l)} = \underline{\underline{F}}_{\backslash l}(\underline{x} - \underline{\tilde{x}}^{(\eta,l)}) \tag{11}$$

as a random vector having a multivariate Gaussian distribution with correlation matrix $\underline{\underline{\Phi}}_{\underline{i}\,\underline{i}}^{(\eta,l)}$. As already stated, the derivation of the soft decision function is based on minimizing the MSE of the decision $\underline{\tilde{x}}_l^{(\eta)}$:

$$\underline{\tilde{x}}_l^{(\eta)} = \arg\left(\min_{\underline{\tilde{x}}_l} \mathrm{E}\{(\underline{x}_l - \underline{\tilde{x}}_l)^T(\underline{x}_l - \underline{\tilde{x}}_l)|\underline{\xi}^{(\eta,l)}\}\right) \tag{12}$$

The solution to this minimization problem is given by:

$$\begin{aligned} \underline{\tilde{x}}_l^{(\eta)} &= \mathrm{E}\{\underline{x}_l|\underline{\xi}^{\eta,l}\} \\ &= \sum_{j=1}^{M} \underline{a}_j P(\underline{x}_l = \underline{a}_j|\underline{\xi}^{(\eta,l)}) \end{aligned} \tag{13}$$

To evaluate Equation (13) we have to determine $P(\underline{x}_l = \underline{a}_j|\underline{\xi}^{(\eta,l)})$. Applying Bayes' rule $P(\underline{x}_j = \underline{a}_m|\underline{\xi}^{(\eta,l)})$ transforms to:

$$P(\underline{x}_l = \underline{a}_j|\underline{\xi}^{(\eta,l)}) = \frac{p(\underline{\xi}^{(\eta,l)}|\underline{x}_l = \underline{a}_j)P(\underline{x}_l = \underline{a}_j)}{p(\underline{\xi}^{(\eta,l)})}\,. \tag{14}$$

The additional assumption of equiprobable transmit symbols, i. e. , $P(\underline{x}_l = \underline{a}_m) = 1/M \;\forall\, \underline{a}_m \in \mathcal{A}$, combined with Equation (14) leads to:

$$P(\underline{x}_l = \underline{a}_j|\underline{\xi}^{(\eta,l)}) = \frac{p(\underline{\xi}^{(\eta,l)}|\underline{x}_l = \underline{a}_j)}{\sum_{m=1}^{M} p(\underline{\xi}^{(\eta,l)}|\underline{x}_l = \underline{a}_m)}\,. \tag{15}$$

For further simplifications we introduce the abbreviation $\underline{w}^{(\eta,l)}$

$$\underline{w}^{(\eta,l)} = \underline{i}^{(\eta,l)} + \underline{n} \tag{16}$$

as the sum of residual interference $\underline{i}^{(\eta,l)}$ plus noise $\underline{n}$. The covariance matrix of $\underline{w}^{(\eta,l)}$ is given by:

$$\underline{\underline{\Phi}}_{\underline{w}\,\underline{w}}^{(\eta,l)} = \underline{\underline{\Phi}}_{\underline{i}\,\underline{i}}^{(\eta,l)} + \sigma_n^2 \underline{\underline{I}} \tag{17}$$

Additionally exploiting the assumption of having interference with Gaussian

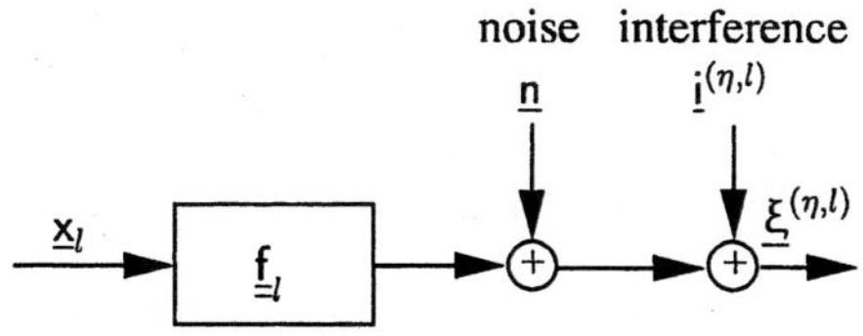

Figure 5. Model of the decision of the SCE

distribution $p(\underline{\xi}^{(\eta,l)}|\underline{x}_l = \underline{a}_j)$ is described by a $2N$ dimensional Gaussian distribution with variance $\underline{\underline{\Phi}}_{\underline{w}\,\underline{w}}^{(\eta,l)}$ and mean $\underline{\underline{f}}_l \underline{a}_j$

$$p(\underline{\xi}^{(\eta,l)}|\underline{x}_l = \underline{a}_j) = \frac{1}{(2\pi)^N \det(\underline{\underline{\Phi}}_{\underline{w}\,\underline{w}}^{(\eta,l)})^{1/2}} \cdot \exp\left(-\frac{1}{2}(\underline{\xi}^{(\eta,l)} - \underline{\underline{f}}_l\underline{a}_j)^T {\underline{\underline{\Phi}}_{\underline{w}\,\underline{w}}^{(\eta,l)}}^{-1} (\underline{\xi}^{(\eta,l)} - \underline{\underline{f}}_l\underline{a}_j)\right) . \quad (18)$$

To reduce the computational complexity it is quite common to use a set of sufficient statistics $\underline{\tilde{\xi}}^{(\eta,l)}$ of the random process, which still contains all information of $\underline{\xi}^{(\eta,l)}$, but requires less dimensions. The condition for a sufficient statistic is summarized in Equation (19).

$$p(\underline{\xi}^{(\eta,l)}|\underline{x}_l = \underline{a}_j) = p(\underline{\tilde{\xi}}^{(\eta,l)}|\underline{x}_l = \underline{a}_j) f(\underline{\xi}^{(\eta,l)}) \quad (19)$$

Then, as it can be easily deduced from Equation (15)

$$P(\underline{x}_l = \underline{a}_j|\underline{\xi}^{(\eta,l)}) = P(\underline{x}_l = \underline{a}_j|\underline{\tilde{\xi}}^{(\eta,l)}) \quad (20)$$

has to hold.

A quite well-known method in the presence of colored noise, which fulfills Equation (19) is the whitening approach. It is composed of a whitening filter and a consecutive matched filter.

$$\underline{\tilde{\xi}}^{(\eta,l)} = (\underline{\underline{f}}_l^T {\underline{\underline{\Phi}}_{\underline{w}\,\underline{w}}^{(\eta,l)}}^{-1} \underline{\underline{f}}_l)^{-1} \underline{\underline{f}}_l^T {\underline{\underline{\Phi}}_{\underline{w}\,\underline{w}}^{(\eta,l)}}^{-1} \underline{\xi}^{(\eta,l)} \quad (21)$$

Application of this approach permits to derive a sufficient statistic $\underline{\tilde{\xi}}^{(\eta,l)}$ of $\underline{\xi}^{(\eta,l)}$ of length 2. It has mean $\underline{a}_j$ and variance $(\underline{\underline{f}}_l^T {\underline{\underline{\Phi}}_{\underline{w}\,\underline{w}}^{(\eta,l)}}^{-1} \underline{\underline{f}}_l)^{-1}$

$$p(\underline{\tilde{\xi}}^{(\eta,l)}|\underline{x}_l = \underline{a}_j) = \frac{\det(\underline{\underline{f}}_l^T {\underline{\underline{\Phi}}_{\underline{w}\,\underline{w}}^{(\eta,l)}}^{-1} \underline{\underline{f}}_l)^{\frac{1}{2}}}{2\pi} \cdot \exp(-\frac{1}{2}(\underline{\tilde{\xi}}^{(\eta,l)} - \underline{a}_j)^T \underline{\underline{f}}_l^T {\underline{\underline{\Phi}}_{\underline{w}\,\underline{w}}^{(\eta,l)}}^{-1} \underline{\underline{f}}_l (\underline{\tilde{\xi}}^{(\eta,l)} - \underline{a}_j)) \quad (22)$$

and can be used in Equation (15) instead of $p(\underline{\xi}^{(\eta,l)}|\underline{x}_l = \underline{a}_j)$.

3.2.4 Estimation of the Covariance matrix

The covariance matrix of the interference may be expressed in terms of a conditional expectation value as shown in Equation (23).

$$\begin{aligned}\underline{\underline{\Phi}}^{(\eta,l)}_{\underline{i}\,\underline{i}} &= \mathrm{E}\{\underline{i}^{(\eta,l)}\,\underline{i}^{(\eta,l)^T}|\underline{\xi}\} \\ &= \underline{\underline{F}}_{\backslash l}\mathrm{E}\{(\underline{x}-\tilde{\underline{x}}^{(\eta,l)})(\underline{x}-\tilde{\underline{x}}^{(\eta,l)})^T|\underline{\xi}\}\underline{\underline{F}}^T_{\backslash l} \\ &= \underline{\underline{F}}_{\backslash l}\underline{\underline{\Phi}}^{(\eta,l)}_{\underline{x}\,\underline{x}}\underline{\underline{F}}^T_{\backslash l}\end{aligned} \tag{23}$$

It depends on the matrix $\underline{\underline{F}}_{\backslash l}$ and the covariance matrix of the remaining decision error $\underline{\underline{\Phi}}^{(\eta,l)}_{\underline{x}\,\underline{x}}$. Under the assumption of statistical independent decision errors of the symbols in one block $\underline{\underline{\Phi}}^{(\eta,l)}_{\underline{x}\,\underline{x}}$ is a block diagonal matrix

$$\underline{\underline{\Phi}}^{(\eta,l)}_{\underline{x}\,\underline{x}} = \underline{\underline{\mathrm{diag}}}([\underline{\underline{\Phi}}^{(\eta-1)}_{\underline{x}_1\underline{x}_1},\dots,\underline{\underline{\Phi}}^{(\eta-1)}_{\underline{x}_l\underline{x}_l},\underline{\underline{\Phi}}^{(\eta)}_{\underline{x}_{l+1}\underline{x}_{l+1}},\dots,\underline{\underline{\Phi}}^{(\eta)}_{\underline{x}_N\underline{x}_N}]) \tag{24}$$

with 2×2 sub-matrices

$$\underline{\underline{\Phi}}^{(\eta)}_{\underline{x}_l\underline{x}_l} = \mathrm{E}\{\underline{x}_l\underline{x}_l^T|\underline{\xi}^{(\eta,l)}\} - \tilde{\underline{x}}_l^{(\eta)}\tilde{\underline{x}}_l^{(\eta)^T}\,. \tag{25}$$

These sub-matrices describe the decision error of the l-th symbol which can be calculated using Equation (25) together with Equation (15)

$$\mathrm{E}\{\underline{x}_l\underline{x}_l^T|\underline{\xi}^{(\eta,l)}\} = \sum_{j=1}^{M}\underline{a}_j\underline{a}_j^T P(\underline{x}_l = \underline{a}_j|\underline{\xi}^{(\eta,l)})\,. \tag{26}$$

3.2.5 Complexity Reduction

The task requiring most computational complexity in the equalization scheme is the matrix inversion of the covariance matrix $\underline{\underline{\Phi}}^{(\eta,l)}_{\underline{w}\,\underline{w}}$. A simplification which leads for many scenarios to a negligible loss is the assumption of having mutually uncorrelated interference terms but taking the correlation of the quadrature components of each interfering term into account. In this case $\underline{\underline{\Phi}}^{(\eta,l)}_{\underline{i}\,\underline{i}}$ and as a consequence of this $\underline{\underline{\Phi}}^{(\eta,l)}_{\underline{w}\,\underline{w}}$ is approximated by a block diagonal matrix with blocks of 2×2 sub-matrices. This simplification reduces the complexity from inverting a $2N \times 2N$ matrix to the complexity of inverting N 2×2 matrices.

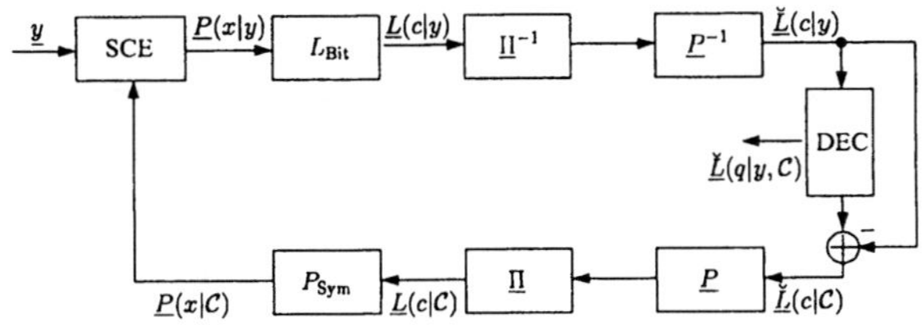

Figure 6. Combined Equalization and Decoding

When additionally neglecting the correlation between the quadrature components of the interference vector $\underline{i}^{(\eta,l)}$, $\underline{\underline{\Phi}}_{\mathrm{ii}}^{(\eta,l)}$ is given by a diagonal matrix:

$$\underline{\underline{\Phi}}_{\mathrm{ii}}^{(\eta,l)} \approx \underline{\underline{\mathrm{diag}}}(\underline{\mathrm{diag}}(\underline{\underline{\Phi}}_{\mathrm{ii}}^{(\eta,l)})) \tag{27}$$

This simplifying approach was pursued in [3].

4. JOINT EQUALIZATION AND DECODING

In the previous sections we have described an iterative equalizer based on the optimum receive filter in the presence of colored Gaussian noise. In this section we explain how to combine equalization and decoding. In the literature this receiver structure is known as "Turbo" equalization scheme. It was first introduced in [2]. In this scheme a combined equalization and decoding block with an unrealizable high complexity is approximated by a structure using two separated devices passing information between each other.

A common probability measure in iterative schemes for binary random variables, e. g. , code bits, are the log likelihood ratio (LLR) L:

$$L(x) = \ln \frac{P(x=0)}{P(x=1)} \tag{28}$$

A receiver structure using the ideas mentioned above is shown in Figure 6. For better readability the indices in the L values and probability arguments have been omitted.

First we describe the exchange of information between decoder and equalizer. To take the constraint of the code into account we modify the probability $P(\underline{x}_l = \mathsf{a}_j | \underline{\xi}^{(\eta,l)})$ in Equation (13) and (26) by adding an additional condition C which represents the constraint of the code. Applying Bayes' rule results in:

$$P(\underline{x}_l = \mathsf{a}_j | \underline{\tilde{\xi}}^{(\eta,l)}, C) = \frac{p(\underline{\tilde{\xi}}^{(\eta,l)} | \underline{x}_l = \mathsf{a}_j, C) P(\underline{x}_l = \mathsf{a}_j | C)}{p(\underline{\tilde{\xi}}^{(\eta,l)} | C)} \tag{29}$$

The additional assumption of statistical independent constraints $\underline{\xi}^{(\eta,l)}$ and $\mathcal{C}$, which is at least justified in the first iteration of the combined equalization decoding scheme by the usage of the random interleaver $\underline{\underline{\Pi}}$, leads to:

$$P(\underline{x}_l = \underline{a}_j | \underline{\tilde{\xi}}^{(\eta,l)}, \mathcal{C}) = \frac{p(\underline{\tilde{\xi}}^{(\eta,l)} | \underline{x}_l = \underline{a}_j) P(\underline{x}_l = \underline{a}_j | \mathcal{C})}{p(\underline{\tilde{\xi}}^{(\eta,l)})} \quad (30)$$

With $\mathcal{B}$ as the mapping from symbols to bits, e. g. , $\mathcal{B}(\underline{a}_1, 1)$ is is the bit value of the first bit of symbol $\underline{a}_1$, and assuming that the additional information $P(c_{li}|\mathcal{C})$ of the code bits determining the l-th symbol is statistical independent $P(\underline{x}_l = \underline{a}_j|\mathcal{C})$ is given by:

$$P(\underline{x}_l = \underline{a}_j | \mathcal{C}) = \prod_{i=1}^{\log_2(M)} P(c_{li} = \mathcal{B}(\underline{a}_j, i) | \mathcal{C}) \quad (31)$$

Using LLR's Equation (31) can be reformulated as

$$P(\underline{x}_l = \underline{a}_j | \mathcal{C}) = \prod_{i=1}^{\log_2(M)} \frac{\exp((1 - \mathcal{B}(\underline{a}_j, i)) L(c_{li}|\mathcal{C}))}{(1 + \exp(L(c_{li}|\mathcal{C})))} \ . \quad (32)$$

Up to now we have described the exchange of information between decoder and equalizer. The exchange between equalizer and decoder is accomplished by a block L_{Bit} which calculates Bit LLR's $L(c_{li}|\underline{y})$ based on the symbol probabilities $P(\underline{x}_l|\underline{\xi}^{(\eta,l)})$ of the equalizer.

$$L(c_{li}|\underline{y}) = L(c_{li}|\underline{\tilde{\xi}}^{(\eta,l)}) = \ln \frac{\sum\limits_{\underline{a}_j \in \mathcal{A}_i^{[0]}} P(\underline{x}_l = \underline{a}_j | \underline{\tilde{\xi}}^{(\eta,l)})}{\sum\limits_{\underline{a}_j \in \mathcal{A}_i^{[1]}} P(\underline{x}_l = \underline{a}_j | \underline{\tilde{\xi}}^{(\eta,l)})} \quad (33)$$

Herein, $\mathcal{A}_i^{[0]}$ denotes the set of modulation symbols which are defined by a bit sequence having a zero at position i whereas $\mathcal{A}_i^{[1]}$ corresponds to the set of modulation symbols having a one at position i.

Several strategies how to pass the information between equalizer and decoder are known in the literature. The one which we pursue in the simulations is a block-wise update per iteration, i. e. , per iteration the information between equalizer and decoder and vice versa is exchanged once.

5. SIMULATION RESULTS

To investigate the performance of the combined equalization and decoding scheme we have computed bit error rates (BER) based on Monte Carlo simulations. We have chosen time invariant channel models according to [8].

Table 1. Symbol time spaced channel models according to [8]

Channel b	$\underline{h} = [0.407\ 0815\ 0.407]$
Channel c	$\underline{h} = [0.227\ 0.460\ 0.688\ 0.460\ 0.227]$

In the coded case we have used a constraint length 3 rate $\frac{1}{2}$ convolutional code with generator polynomial [7 5]. The blocklength of the terminated convolutional code was adjusted such that the corresponding bit sequence of K equalizer blocks is determined by one code word of length N_c. The corresponding parameter set is addressed by $K \times N$.

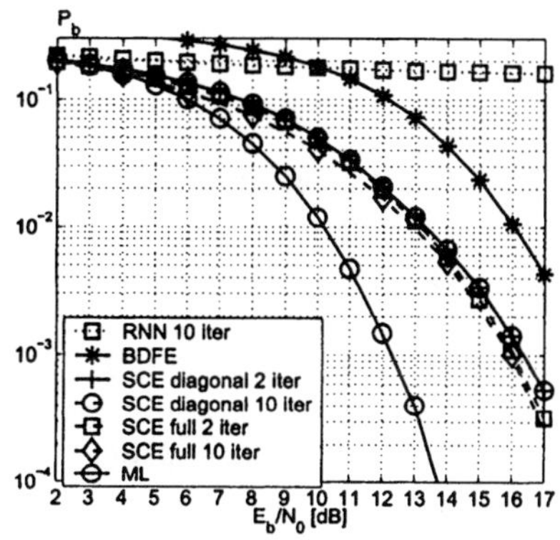

Figure 7. Comparison between SCE using diagonal and full covariance matrix, BDFE and RNN, BPSK, channel c, blocklength $N = 16$

Figure 7 shows the results obtained for uncoded BPSK, an equalizer blocklength $N = 16$ and channel c. As an upper bound the classical CBDFE is depicted whereas maximum likelihood (ML) detection serves as a lower bound. In this case the full complexity and the diagonal approximation of the SCE perform nearly the same. Furthermore, the improvement when using more than 2 iterations in negligible. The performance gain at a BER of 10^{-1} of 3.5 dB of the SCE in comparison to the CBDFE as well as the performance loss in comparison to ML of about 1.5 dB is remarkable. Furthermore, the RNN type equalizer does not converge at all.

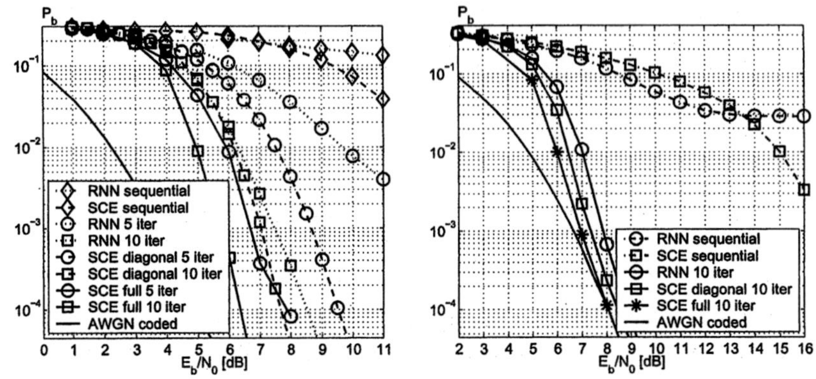

(a) BPSK, channel c

(b) 16 QAM, channel b

Figure 8. Comparison between SCE using diagonal and full covariance matrix and RNN, Blocksize 100x16

The results for coded BPSK transmission using a blocklength 100×16 and channel c are shown in Figure 8(a). As a lower bound the code performance in an AWGN channel without ISI is depicted. Besides the curves for combined equalization and decoding we have added curves for sequential equalization and decoding, i. e. , there is no iteration process between equalization and decoding. In the graphs these curves are labeled with sequential and show the performance for the SCE with diagonal approximation and the RNN using 10 iterations. Comparing the different curves, the first thing we notice is the impressive improvement of the combined schemes in comparison to the sequential ones. Furthermore, we see that the SCE with full complexity nearly achieves the lower bound and outperforms the SCE with diagonal approximation. The RNN which did fail in the uncoded case offers in the coded case moderate performance. We note that an increased of number of iterations up to a number of 10 improves the performance significantly. Figure 8(b) shows the corresponding results for coded 16 QAM, channel b, and 100x16 blocks. Again the improvement of the combined schemes compared to the corresponding sequential ones is impressive. For this scenario, all combined schemes yield good performance and nearly achieve the lower coded AWGN bound. Comparing the combined schemes in more detail, we see that as expected the SCE with full complexity performs best followed by the SCE with diagonal approximations and the RNN.

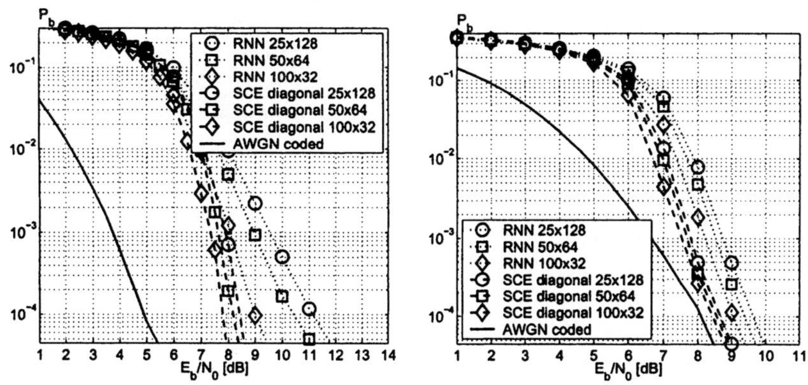

(a) BPSK, channel c (b) 16 QAM, channel b

Figure 9. Comparison between SCE using diagonal approximation and RNN for varying blocklength using 10 iterations

Results for varying equalizer blocklength while keeping the codeword length constant are shown for the BPSK and channel c case in Figure 9(a) and for the 16 QAM and channel b case in Figure 9(b) . Both equalizers in both scenarios suffer from a performance degradation for increasing equalizer blocklength. Nevertheless, the decrease is quite moderate for both equalizers especially for the SCE.

6. CONCLUSIONS

A new iterative equalization and decoding scheme based on the classical CBDFE has been introduced. It has been shown by simulation that this combination, even under some simplifying assumptions which reduce complexity, can achieve near optimum performance.

REFERENCES

[1] A. Bury and J. Lindner, "Block transmission equalizers using constrained mininumum variance filters with application to mc-cdm ", In *IEEE 6th International Symposium on Spread-Sprectrum Techniques and Applications, 6.9.-8.9.2000, New Jersey*, pages 159–163, Sept 1999.

[2] C. Douillard, M. Jezequel, C. Berrou, A. Picart, P. Didier, and A. Glavieux, "Iterative Correction of Intersymbol Interference: Turbo-Equalization". *European Transactions on Telecommunications*, 6(5):507–511, Sept.-Oct. 1995.

[3] J. Egle, C. Sgraja, and J. Lindner, "Iteratitive soft cholesky block decision feedback equalizer – a promising approach to combat interference", In *Vehicular Technology Conference 2001 (spring)*, pages 1604–1608, 2001.

[4] T. Frey and M. Reinhardt, "Signal Estimation for Interference Cancellation and Decision Feedback Equalization", In *Vehicular Technology Conference 1997*, pages 155–159, 1997.

[5] A. Lampe, R. Schober, and W. Gerstacker, "A novel iterative multiuser detector for complex modulation schemes", In *Proceedings COST 262 Workhop Multiuser Detection in Spread Spectrum Communications*, pages 25–29, 2001.

[6] J. Lindner, "MC-CDMA in the Context of General Multiuser / Multisubchannel Transmission methods". *European Transactions on Telecommunications*, 10:351–367, 1999.

[7] F. D. Neeser and J. L. Massey, "Proper complex random processes with application to information theory", *IEEE Transactions on Information Theory*, vol. 39(no. 4):pp. 1293–1302, 1994.

[8] J. G. Proakis, *Digital Communications*, McGraw-Hill, 1989.

[9] M. C. Reed, *Iterative Receiver Techniques for Code Multiple Access Communication Systems*, Ph.d. theses, University of South Australia, Faculty of Information Technology, 1998.

[10] C. Sgraja, A. Engelhart, W. G. Teich, and J. Lindner, "Combined equalization and decoding for BFDM packet transmission schemes", In *Proceedings 1st international OFDM-Workshop, 22./23. Sept.*, 1999.

[11] C. Sgraja, A. Engelhart, W. G. Teich, and J. Lindner, "Equalization with recurrent neural networks for complex-valued modulation schemes", In *Proc. 3rd Workshop Kommunikationstechnik, Schloß Reisensburg/Germany*, 1999.

[12] F. Tarköy, "MMSE-optimal feedback and its applications", In *IEEE International Symposium on Information Theory, Whistler, B.C., Canada*, page 195, 1995.

[13] W. G. Teich and M. Seidl, "Code Division Multiple Access Communications: Multiuser Detection based on a Recurrent Neural Network Structure", In *International Symposium on Spread Spectrum Techniques and Applications, 22.9.-25.9.1996, Mainz, Germany*, pages 979–984, 1996.

INDEX

Kolibri 360 GmbH
Ferdinand-Jühlke-Straße 7
99095 Erfurt, Deutschland
produktsicherheit@kolibri360.de